COMMITTEE ON FISH STOCK ASSESSMENT METHODS

RICHARD DERISO *(Cochairman),* Inter-American Tropical Tuna Commission, La Jolla, California
TERRANCE QUINN *(Cochairman),* University of Alaska Fairbanks, Juneau
JEREMY COLLIE, University of Rhode Island, Narragansett
RAY HILBORN, University of Washington, Seattle
CYNTHIA JONES, Old Dominion University, Norfolk, Virginia
BRUCE LINDSAY, Pennsylvania State University, University Park
ANA PARMA, International Pacific Halibut Commission, Seattle, Washington
SAUL SAILA, University of Rhode Island, Narragansett
LYNDA SHAPIRO, University of Oregon, Charleston
STEPHEN JOSEPH SMITH, Bedford Institute of Oceanography, Dartmouth, Nova Scotia, Canada
CARL WALTERS, University of British Columbia, Vancouver, Canada

Staff

EDWARD R. URBAN, JR., Study Director
LORA TAYLOR, Senior Project Assistant

Preface

Global marine fish harvest has plateaued and many important commercial stocks have been depleted. The Ocean Studies Board (OSB), at the request of the National Marine Fisheries Service (NMFS), has provided advice designed to improve management of marine fisheries in the United States (NRC, 1994a). Many of the changes suggested in the 1994 report *Improving the Management of U.S. Marine Fisheries* were incorporated in the reauthorized Magnuson-Stevens Fishery Conservation and Management Act (MSFCMA) in 1996.

NMFS also has asked the OSB for advice on specific issues of Atlantic bluefin tuna population biology; the OSB presented its results in another 1994 report, *An Assessment of Atlantic Bluefin Tuna* (NRC, 1994b). Following the publication of that report, there was a widespread expression of the need for reviews of several other specific stock assessments. NMFS responded by requesting a broad review of the methods used in the United States for stock assessments. The results of that review are presented in this report.

This study would not have been possible without the efforts of NMFS scientists who carried out blind runs of data provided by the study committee and participated in the committee's meetings. The cooperation between academic and agency scientists was commendable and actually led to advances in the state of the art of fish stock assessments. The OSB offers sincere thanks to NMFS analysts for their considerable efforts, including Ray Conser, Jeff Fujioka, Wendy Gabriel, Phil Goodyear, Jim Ianelli, Rick Methot, Jerry Pella, Clay Porch, Joe Powers, Mike Prager, Victor Restrepo, Gerald Scott, and Mike Sigler. Other individuals also contributed to the committee's work and deserve the thanks of the committee and the OSB: Jie Zheng (Alaska Department of Fish and Game), Andre Punt (Commonwealth Scientific and Industrial Research Organization, Australia), and David Fournier (Otter Research Ltd.).

The results of this study will serve as an important foundation for the National Research Council review of the U.S. Northeast fisheries stock assessments that was mandated by Congress in 1996 as part of the MSFCMA reauthorization. The recommendations presented herein should also be useful to ongoing international activities related to fish stock assessments, such as those of the International Council for the Exploration of the Sea.

Kenneth Brink
Chairman, Ocean Studies Board

Contents

Executive Summary

Marine fisheries provide a vital contribution to food supplies, employment, and culture worldwide. Therefore, matching fishing activities with natural fluctuations so as to avoid unsustainable harvests and population crashes is an important goal. In an ideal world, accurate and precise estimates of the abundance of fish stocks and their dynamics (how and why population levels change) would be available to set sustainable harvest levels to accommodate commercial and recreational demand. In reality, fishery management is based on imperfect estimation of the number, biomass, productivity, and age structure of fish populations and incomplete knowledge of population dynamics. The ocean is relatively opaque to light, and acoustic techniques of remote sensing are not yet sufficiently developed for general use in estimating fish populations. Thus, it is difficult to count fish through nondestructive means and fish usually must be caught to be counted, weighed, and measured. Standardized techniques have been developed to sample a relatively small proportion of fish from a population and to combine such data with commercial and recreational catch information to estimate population characteristics. These techniques yield *stock assessments* used by managers at state, regional, national, and international levels.

In addition to monitoring the abundance and productivity of exploited fish populations, stock assessments can provide a quantitative prediction of the consequences of possible alternative management actions. The mechanisms that cause fish populations to change are poorly understood but include environmental and ecosystem effects, interactions among multiple species, and effects of humans through harvesting, pollution, habitat disruption, and other factors. Without accurate stock assessments and their proper use in management, exploited fish populations can collapse, creating severe economic, social, and ecological problems. Therefore, ensuring that stock assessment research progresses and that operational stock assessments use the best techniques for a given stock are fundamental for ensuring the sustainability of commercial and recreational marine fisheries.

Stock assessment is a multistage process. Steps include (1) definition of the geographic and biological extent of the stock, (2) choice of data collection procedures and collection of data, (3) choice of an assessment model and its parameters and conduct of assessments, (4) specification of performance indicators and evaluation of alternative actions, and (5) presentation of results. This report concentrates on evaluating assessment models, with less extensive treatment of the other steps. Chapter 1 discusses these steps in greater detail. Techniques of stock assessment range from informal estimates to more sophisticated modeling approaches used to combine data of various types. Assessment models predict rates of change in biomass and productivity based on information about yield from fisheries and the rates at which fish enter the harvestable population (*recruitment*), grow in size, and exit the population (*natural* and *fishing mortality*).

Stock assessments for fish living in the U.S. exclusive economic zone (3 to 200 nautical miles from shore) and for some highly migratory species are conducted by scientists from the National Oceanic and Atmospheric Administration's (NOAA's) National Marine Fisheries Service (NMFS) and independent species group commissions (e.g., the International Pacific Halibut Commission and the Inter-American Tropical Tuna Commission). In addition, interstate fishery management commissions were created to facilitate the coordination of state assessment scientists in working with each other and with federal scientists to assess and manage stocks shared among states in their coastal waters (within 3 nautical miles from shore on open coasts, as well as bays and estuaries). These organizations include the Atlantic States, Gulf States, and Pacific States Marine Fisheries Commissions. Some states (e.g., Alaska, Oregon, and Florida) also perform assessments for fisheries conducted in their own state waters.

Fishery management organizations use the results of stock assessments to design and implement various controls for the total catch that can be removed from fish populations under their jurisdictions. Commercial catch can be managed by specifying the amount of harvesting allowed; the areas of fishing and times of the year that fishing can take place; the gear that can be used; minimum fish size limits; and in some cases, the amount of fish that any single fisher, community, company, or other entity can catch. Recreational fisheries more often impose minimum size limits, daily catch limits, seasons, and sometimes gear restrictions and requirements to release fish that are caught.

STUDY PROCESS

The National Research Council (NRC) Committee on Fish Stock Assessment Methods was formed in early 1996 to review existing stock assessment methods and to consider alternative approaches for the future. The committee's statement of task was two-fold:

1. Conduct a scientific review of stock assessment methods and models for marine fisheries management.
2. Compare models using actual and simulated data having a variety of characteristics, to test the sensitivity and robustness of the models to data quality and type.

As part of this study, the committee asked selected stock assessment scientists to conduct blind runs of simulated data sets using five different models. Models tested included a production model, a delay-difference model, and three age-structured methods (described in detail in Chapter 3). The goal of the simulation study was to evaluate the performance of stock assessment methods for simulated fish populations for which the true population parameters were known (to the committee, but not to the analysts) and some of the assumptions usually made in stock assessments were violated. One type of data set was typical of the catch biomass, age composition of the catch, and catch per unit effort (CPUE) that are obtained from commercial and recreational fisheries. The other type of data set was typical of that collected by fishery-independent surveys.

Each analyst was asked to evaluate five 30-year sets of simulated commercial and survey data, alone and in combination. The five data sets provided different combinations of parameters in terms of the following:

- Increasing or decreasing stock size over time (*population trend*)
- Constant versus changing age of fish caught (*fishery selectivity*) over time
- Accuracy of catch reported by fishers
- Ability of fishery and survey vessels to catch fish (*fishery* and *survey catchability*)

The analysts were given essential information about fish growth and maturity, the probability of mis-estimating fish ages, and selected information about the structure of the populations and the data. Analysts were not provided information about natural mortality, catchability, selectivity, recruitment, or the amount of underreporting (although they were warned that underreporting might have occurred).

In addition to the results of these basic analyses, (1) some analysts repeated their model runs with the true average natural mortality (provided by the committee), (2) key management variables were calculated by analysts and the committee, and (3) retrospective analyses were conducted by the committee to determine the persistence of over- or underestimation of population parameters over time by the different models. Greater detail about the study process is given in Chapter 5 and Appendix E.

FINDINGS AND RECOMMENDATIONS

The committee focused its examination on the data that are used in assessments, model performance, use of harvest strategies, new assessment techniques, periodic review and quality control of assessments and assessment methods, and education and training of stock assessment scientists. The committee based its recommendations on the results of the simulations and on its collective experience. Caveats about how the analyses conducted for this study compare to actual stock assessments are given in Chapter 5. Accomplishing the recommendations of this report will require concerted and cooperative action by all interested parties (academic and government scientists, fishery managers, user groups, and environmental nongovernmental organizations) to improve the stock assessment process and products.

Data Collection and Assessment Methods

The committee concludes that stock assessments do not always provide enough information to evaluate data quality and to estimate model parameters, and it recommends a checklist that would promote more complete data collection for use in stock assessments. The results of the committee's simulations demonstrated that the availability of continuous sets of data collected by using standardized and calibrated methods is important for the use of existing stock assessment models. The best index of fish abundance is one for which extraneous influences (e.g., changes in gear and seasonal coverage, changes in fishers' behavior) can be controlled. The committee recommends that at least one reliable abundance index should be available for each significant stock. CPUE data from commercial fisheries, if not properly standardized, do not usually provide the most appropriate index. Likewise, CPUE data from recreational fisheries require standardization to serve as a good index of abundance.

Fishery-independent surveys offer the best opportunity for controlling sampling conditions over time and the best choice for achieving a reliable index if they are designed well with respect to location, timing, sampling gear, and other considerations of statistically valid survey design. NMFS should support the long-term collection of fishery-independent data, using either the NOAA fleet or calibrated independent vessels. Diminishing the quality of fishery-independent data by failing to modernize NOAA fishery research vessels or by changing sampling methods and gear without proper calibration could reduce the usefulness of existing and future data sets.

The simulation study demonstrated that assessments are sensitive to underlying structural features of fish stocks and fishery practices, such as natural mortality, age selectivity, catch reporting, and variations in these or other quantities. Auxiliary information in the form of indices or survey estimates of abundance, population structure information, and accurate estimates of other population parameters (e.g., natural or fishing mortality, growth, catchability) improves the accuracy of assessments.

Formally reviewed sampling protocols for collection of commercial fisheries statistics have not been implemented in many geographic regions. The lack of formalized, peer-reviewed data collection methods in commercial fisheries is problematic because bias and improper survey conduct may exist, with unknown impact on data reliability. Greater attention should be devoted to sampling design based on an understanding of the statistical properties of the estimators for catch at age and other factors. Sampling and subsequent analysis should also consider the issue of systematic biases that emerge with factors such as misreporting. Formalized sampling protocols have been developed for recreational fisheries in the form of the Marine Recreational Fisheries Statistics Survey (MRFSS). MRFSS data and methods, albeit imperfect, have undergone independent peer review, are readily available, and could serve as a model for commercial fisheries. The committee recommends that a standardized and formalized data collection protocol be established for commercial fisheries nationwide.

Models

Both harvesting strategies and decision rules for regulatory actions have to be evaluated simultaneously to determine their combined ability to sustain stocks. Simulation models should be realistic and encompass a wide range of possible stock responses to management actions and natural fluctuations consistent with experience. The committee recommends that fish stock assessments present realistic measures of the uncertainty in model outputs whenever feasible. Although a simple model can be a useful management tool, more complex models are needed to better quantify the unknown aspects of the system and to address the long-term consequences of specific decision rules adequately. Retrospective analyses performed by the committee showed that persistent over- or underestimation can occur over a number of years of assessment, regardless of which model is used. The committee recommends the use of Bayesian methods both for creating distributions of input variables and for evaluating alternative management policies. Other methods for including realistic levels of uncertainty in models also should be investigated.

In the simulations, model performance became erratic as more variability or errors were introduced to data sets. Newer modeling methods offer promise for reducing bias in key parameter estimates, although using mathematically sophisticated assessment models did not mitigate poor data quality. Different assessment models should be used to analyze the same data to help recognize poor data and to improve the quality of assessment results. Results from such comparisons can be used to direct survey programs to improve data quality and to assess the degree of improvement in data achieved over time. Greater attention should also be devoted to including independent estimates of natural mortality and its variability in assessment models. Further simulation work of this kind is also needed to determine whether the simulation results and the conclusions based on these results remain the same over multiple replications.

The committee believes that single-species assessments provide the best approach at present for assessing population parameters and providing short-term forecasting and management advice. Recent interest in bringing ecological and environmental considerations and multi-species interactions into stock assessments should be encouraged, but not at the expense of a reduction in the quality of stock assessments.

Harvest Strategies

Although the committee did not evaluate alternative harvest strategies, it believes that assessment methods and harvest strategies should be evaluated together because harvest strategies can affect stock assessments and the uncertainty inherent in stock assessments should be reflected in harvest strategies. Despite the uncertainty in stock assessments, fishery scientists may be able to identify robust management measures that can at least prevent overfishing, even if they cannot optimize performance. Conservative management procedures include management tools specific to the species managed, such as minimum biomass levels, size limits, gear restrictions, and area closures (for sedentary species). Management procedures by which the allowable catch is set as a constant fraction of biomass (used for many U.S. fisheries) generally perform better than many alternative procedures. However, errors in implementation due to assessment uncertainties could result in substantial reductions in long-term average harvests in some years if biomass estimates are highly uncertain. Assessment methods and harvest strategies need to be evaluated simultaneously to determine their ability to achieve management goals. Application of risk-adjusted reference points (based on fishing mortality or biomass) would immediately lead to reduced total allowable catch and thus create an economic incentive for investment in improved data gathering and assessment procedures to reduce the coefficient of variation of biomass estimates.

There are at least four alternatives to harvesting a constant fraction of exploitable biomass that may result in levels of total mortality that are consistent with maintaining a fish stock. First, target fishing mortality can be reduced as a stock decreases in size to reduce risks. Second, a minimum biomass level can be established, below which fishing would be halted (this is done for some U.S. fisheries). Third, the size of fish captured can be increased by changing requirements for harvest gear. This restriction might allow smaller fish to escape and spawn, but could be ineffective if harvesters apply more effort to the larger fish. Finally, geographic areas can be closed to limit mortality for sedentary species if the distribution of organisms is well known and if the fishing mortality in other areas is not increased. Area closures have been implemented or proposed for many fisheries worldwide in the form of marine reserves and sanctuaries.

New Approaches

NMFS and other organizations responsible for fisheries management should support the development of new techniques that can better accommodate incomplete and variable data and can account for the effects of environmental fluctuations on fisheries. Such techniques should allow the specification of uncertainty in key parameters (rather than assuming constant, known values), should be robust to measurement error, and should include the ability to show the risks associated with estimated uncertainty.

A few prominent recommendations for new approaches emerged from the study. Scientists that conduct stock assessments and organizations that depend on assessments should

- incorporate Bayesian methods and other techniques to include realistic uncertainty in stock assessment models;
- develop better assessment models for recreational fisheries and methods to evaluate the impacts of the quality of recreational data on stock assessments;
- account for effects of directional changes in environmental variables (e.g., those that would accompany climate change) in new models; and
- develop new means to estimate changes in average catchability, selectivity, and mortality over time, rather than assuming that these parameters remain constant.

The results from the simulation exercise should be sobering to scientists, managers, and the users of fishery resources. The majority of the estimates of exploitable biomass exceeded true values by more than 25%; assessments that used accurate abundance indices performed roughly twice as well as those that use faulty indices. A disturbing feature of the assessment methods is their tendency to lag in their detection of trends in the simulated population abundance over time. For example, some methods with some types of data consistently overestimate exploitable biomass during periods of decreasing simulated abundance and underestimate exploitable biomass during periods of increasing simulated abundance.

Although no stock assessment model was free from significant error in the simulations, it is also true that few of the models failed consistently. Hence, the message of this report is not that stock assessment models should not be used, but rather that data collection, stock assessment techniques, and management procedures need to be improved in terms of their ability to detect and respond to population declines. The simulation results and some actual fishery management examples suggest that overestimation of stock biomass and overfishing of a population can occur due to inaccurate stock assessments and that the overestimation can persist over time. The committee believes that the two most important management actions to mitigate this problem are: (1) to model and express uncertainty in stock assessments explicitly, and (2) to incorporate uncertainty explicitly into management actions such as harvesting strategies.

The absence of adequate data is the primary factor constraining accurate stock assessments. The differences between estimated and true values derived from the simulated data were most likely not introduced by any mistakes made by the analysts. Rather, the large differences that occurred under some scenarios were primarily the result of poor data and model misspecification stemming from incomplete knowledge of the true situation by the analysts. The surplus production and delay difference models did not include the ability to account for changes over time in key parameters for the simulated populations. The simulated data sets were better structured for analysis by age-structured methods; hence, these kinds of models performed better. When they did not perform well, it was generally because the models used biased information (e.g., the fishery CPUE index) or did not account for changes in selectivity and catchability over time. Had the analysts been told about these data features, it is likely that they could have compensated for them and obtained better assessments. Some of the newer models appear to be able to achieve such compensation through the introduction of process errors. Nevertheless, modeling will never be able to provide estimates that are as accurate as direct knowledge obtained by measurement and experimentation. Thus, if future stock assessments are to avoid some of the past problems, management agencies must devote the necessary resources to monitor and investigate fish populations in a stable research environment that fosters creative approaches.

Peer Review

It is imperative that stock assessment procedures and results be understood better and trusted more by all stakeholders. One means to achieve such trust is to conduct independent peer review of fishery management methods and results including (1) the survey sampling methods used in data collection, (2) stock assessment procedures, and (3) risk assessment and management strategies. When applied properly to stock assessments, peer review yields an impartial evaluation of the quality of assessments as well as constructive suggestions for improvement. Such reviews are most beneficial when conducted periodically, for example, every 5 to 10 years, as new information and practices develop. In addition, a complete review of methods for collection of data from commercial fisheries should be conducted in the near future by an independent panel of experts, which could lead to the adoption of formal protocols.

Education and Training

Reduction in the supply of stock assessment scientists would endanger the conduct of fishery assessments by the federal government, interstate commissions, and international management organizations and would hinder progress in the development and implementation of new stock assessment methods. NMFS and other bodies that conduct and depend on fish stock assessments should cooperate to ensure a steady supply of well-trained stock assessment scientists by using mechanisms such as personnel exchanges among universities, government laboratories, and industry and by funding stock assessment research activities. The training of stock assessment scientists should endow them with skills in applied mathematics, fisheries biology, and oceanography. Education of fisheries scientists should be organized and executed in such a way that it complements and augments the NMFS research mission and leads to improved management strategies for fisheries in the future.

1

Introduction

Catches from fish populations managed within U.S. waters have fluctuated substantially over the last century and in many cases have declined precipitously, creating serious economic, social, and ecological problems. Accurate assessments of population characteristics (such as mean and variance of estimates of annual abundance) are crucial for sustaining[*] fisheries while allowing adequate levels of catches. Two primary roles of stock assessment are (1) to monitor the abundance and productivity of exploited fish populations and (2) to provide fishery managers a quantitative evaluation of the potential consequences of alternative actions, to help achieve management goals.

OVERVIEW OF U.S. FISHERIES

The United States has the largest exclusive economic zone (EEZ) of any nation, covering about 11 million square kilometers. The United States ranked fifth in the world in fish harvests in 1993, following China, Japan, Peru, and Chile (FAO, 1995b) and accounts for approximately 7% of the global wild catch of marine fish. The first-sale value of U.S. commercial landings (5.1 million metric tons [mmt]) in 1994 was estimated at $3.8 billion (NMFS, 1995), with a contribution to the U.S. gross national product of $20.2 billion when the extended (*multiplier*) effect on related industries and the national economy is included. The U.S. catch is dominated by a small number of species, with almost 50% of the catch (by weight) composed of walleye pollock (*Theragra chalcogramma*) from Alaskan waters and menhaden (*Brevoortia tyrannus*) from the Gulf of Mexico and the Atlantic Ocean. In terms of harvest value, sockeye salmon (*Oncorhynchus nerka*), walleye pollock, brown shrimp (*Penaeus aztecus*), and American lobster (*Homarus americanus*) accounted for about one-third of the catch in 1995.

Recreational fishing is also important in the United States. Although the recreational catch is only about 2%

---

[*]"Sustainable development has been defined as the management and conservation of the natural resource base, and the orientation of technological and institutional change in such a manner as to ensure the attainment and continued satisfaction of human needs for present and future generations. Such development conserves land, water, plant genetic resources, is environmentally non-degrading, technologically appreciated, economically viable and socially acceptable . . ." (FAO, 1995a, p. 1). Sustainability does not necessarily imply biologic or economic optimality.

as large as commercial landings for all species combined (90,000 metric tons in 1994), there are more than 17 million marine recreational fishers (*anglers*) who make more than 66 million fishing trips per year, catch about 360 million fish, and spend $25.3 billion per year on fishing-related activities (NMFS, 1995a); thus, recreational and commercial fishing activities contribute roughly equally to the U.S. economy. For some fisheries in which both commercial and recreational fishers participate (e.g., summer flounder [*Paralichthys dentatus*] and bluefish [*Pomatomus saltatrix*]), the recreational catch is a significant portion of the total. The allocation of available marine fish resources between commercial and recreational sectors is a major issue for fishery management councils and for state, regional, and federal fisheries agencies.

In its most recent assessment of the condition of U.S. fisheries, the National Marine Fisheries Service (NMFS) evaluated 275 stocks caught by fishers in nearshore coastal waters, the EEZ, and the high seas beyond the EEZ (NMFS, 1996). Of these stocks, there was not enough information available to evaluate the status of 31% of the stocks in 1995; 23% of the stocks were overutilized, 34% were fully utilized, and 12% were underutilized. In 1993, 29% of the stocks were of unknown status, 28% were overutilized, 31% fully utilized, and 12% underutilized (NMFS, 1993). Nearly half the stocks are below the level of abundance that will produce the greatest long-term potential yield (LTPY).* The LTPY of the U.S. fisheries within the U.S. EEZ is estimated to be 8.1 mmt per year, which is more than 60% greater than the recent yield of 5.1 mmt (NMFS, 1996). For the United States to achieve its potential increase in LTPY, currently underutilized fisheries will have to be developed further, but more importantly, fishing on overutilized stocks will have to be reduced so that stocks can rebuild. The estimated LTPY and maximum sustainable yield (MSY) levels can be used as targets to regulate fishing activities to levels that will sustain or rebuild marine fish stocks. Setting of appropriate LTPY and MSY levels depends on accurate stock assessments.

OVERVIEW OF THE USE OF ASSESSMENTS IN MANAGEMENT

Fishery managers are responsible for sustaining fish stocks, and stock assessment scientists are responsible for providing analyses and abundance indices that can make such management possible. There are five steps in a stock assessment (see Appendix D):

1. stock definition,
2. choice of data collection procedures and collection of data,
3. choice of an assessment model and its parameters and conduct of assessments,
4. evaluation of alternative actions and specification of performance indicators, and
5. presentation of results.

Stock Definition

A fish stock can be defined as all fish belonging to a given species that live in a particular geographic area at a particular time, that is, all individuals actually capable of interbreeding. For practical management purposes, a stock is often further defined by political boundaries. That is, the management unit, often still called a stock, includes those members of a biological stock that are under management by a single governmental agency. Units so defined, however, do not necessarily reflect meaningful biological entities or the spatial heterogeneity of fish distributions. When biological stock boundaries extend beyond national jurisdictions, only international cooperation can permit accurate assessments and wise management of the joint resources. The National Research Council

---

*"Long-term potential yield (LTPY) is the maximum long-term average yield that can be achieved through conscientious stewardship, by controlling the fishing mortality rate through regulating fishing effort or total catch. LTPY is a reference point for judging the potential of a resource . . . A fishery resource is considered overutilized when more fishing effort is employed than is necessary to achieve LTPY . . . A fishery resource is classified as underutilized when more fishing effort is required to achieve LTPY" (NMFS, 1996, pp. 151-152).

(NRC, 1994b) provides an example (for Atlantic bluefin tuna) of the implications of stock definitions and discusses a variety of means to delineate stocks.

Even when political considerations do not compromise the definition of a stock, its biological (genetic) boundaries are not always known. Knowledge of such boundaries may not always be vital to proper management of a stock because every breeding individual does not need to be included in a specific stock for management purposes, if the vast majority are included.

To the fishery manager, a stock is a collection of individual fish, of similar morphology and habitat use, that occurs in a certain locality at one time. The collection may or may not have genetic integrity. That is, (1) the managed stock may be a Mendelian population reflecting the genetic concept of a stock; (2) it may contain more than one genetically isolated population; or (3) it may contain only part of a population. The important point is that the stock defined on a management basis is managed as a unit whether or not it is identical to a genetic stock. This introduces a level of uncertainty to demographically based assessment models.

Choice of Data Collection Procedures and Collection of Data

Data regarding the numbers of fish caught, effort expended in catching the fish and other economic data, and biological information about the captured fish are gathered from commercial and recreational fishers (Chapter 2). Catch and yield can be determined from fish tickets, creel surveys, and fishery logbooks. Sampling of landings and/or catch (including discards) provides information regarding number of fish and their length, age, gender, maturity, and fecundity. To verify information obtained from fishers, independent observers can be placed on fishing vessels to record catch, bycatch, discards, and other biological characteristics. Economic performance and efficiency of fisheries can be estimated from effort and catch per unit effort (CPUE); price, cost, and auxiliary variables; and logbooks or interviews. To verify and complement information obtained from fishers, independent surveys are used to estimate biomass, abundance, and other biological characteristics. Survey methods include trawl, longline, pot or trap, mark-recapture, and aerial techniques (Doubleday and Rivard, 1981; Gunderson, 1993). Fishery-independent research can yield other information to aid stock assessments and to improve models. For example, the movement and distribution of fish (determined through tagging and relative CPUE) and the genetic and morphometric features of populations are necessary for determining movement among stocks such as Atlantic bluefin tuna (Gunderson, 1993; NRC, 1994b).

Choice of an Assessment Model and Parameters and Conduct of Stock Assessments

Any stock assessment model involves choices at two levels: the structural model and the parameter values and data to be used (Chapter 3). The purpose of stock assessments is to provide information to fishery managers that will allow them to control the catch of each species or species complex (or the effort directed to it) so that, ideally, populations can be maintained to produce the MSY* or LTPY for each species. Generally, stock assessments estimate the current abundance of a stock, its rate of removal due to fishing, and/or the abundance needed to sustain the stock in the future. Most assessments of U.S. fisheries are conducted by scientists employed by NMFS, although assessments are also conducted by some states, interstate marine fisheries commissions, and in the case of international stocks, international assessment groups.

Stock assessments can be conducted by directly analyzing data characteristics and/or by using models to integrate data. As described below, informal estimates and more formal methods based on abundance are used to manage about half of U.S. marine species (by number of species). More sophisticated modeling approaches are used for the other species, to synthesize data of various types. Different processes affecting changes in stock abundance can be modeled, including recruitment, natural and fishing mortality, gear selectivity and catchability,

---

*Maximum sustainable yield (MSY) is the largest average catch that can be captured from a stock under existing environmental conditions on a sustainable basis.

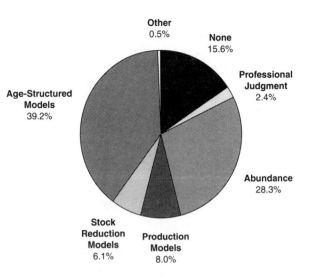

FIGURE 1.1   Stock assessment methods used for U.S. fisheries. The data for this figure were complied in 1995 and were provided to the committee by the National Marine Fisheries Service.

stock-recruitment relationships, and movement rates.   Different assumptions about *process errors** and measurement errors can be used to account for variability in parameters arising from biological or environmental factors and to estimate the extent of such effects.

Stock Assessment Methods Used by NMFS

NMFS surveyed the stock assessment methods used for marine fisheries of the United States and provided summary information to the committee for 212 of the 275 species or complexes it manages. For some fish stocks, no stock assessment is conducted. For those stocks that are assessed, the following methods are used: (1) informal estimates based on professional judgment, (2) abundance analyses, (3) production models, (4) stock reduction models, and (5) age-structured models (Figure 1.1; see Chapter 3 for descriptions of models). For some species, more than one method is used. There is a cost associated with assessments, and there may be some relation of the value of a fishery to the cost expended in conducting stock assessments. Tradition and experience within a region may also influence the method selected.

No Stock Assessment (16%)—No stock assessment is performed for some species, presumably because managers have not requested assessments, the value of the catch does not merit the expense of assessments, and/or the stock is distributed across wide geographic areas. Most of these species are large pelagic fish (e.g., dolphinfish, minor tuna species, some billfish, some sharks) or reef-dwelling fish (e.g., grouper, sea bass, and many forage fish).

Informal Estimates (2%)—Informal estimates based on the professional judgment of assessment scientists are used for a small percentage of managed species, including the Nassau grouper, Atlantic croaker, Pacific jack mackerel, and Caribbean corals.

Analysis of Abundance Trends (28%)—This category of methods includes the use of trends in CPUE and relative abundances over time. CPUE is obtained from fishers. Relative abundances are determined through fishery-independent surveys. For example, swept-area surveys can be used to estimate fish density from the area

---

*Process errors refer to variability in the population dynamics that cannot be accounted for using deterministic population models, but can be modeled as random processes. An example is the relationship between the number of spawning adults and the resulting juveniles entering the population as catchable fish.

sampled by the fishing gear. Some representative species assessed using abundance trends include spiny dogfish, long-finned squid, striped bass, Atlantic shark, and most crustacean species.

Production Models (8%)—These methods include equilibrium, nonequilibrium, and delay-difference models. Representative species include highly migratory species in the Atlantic and Pacific Oceans (for which age-structured sequential population analyses are also conducted), Spanish mackerel, and northern anchovy in the Pacific Ocean.

Stock Reduction or DeLury-Type Models (6%)—Examples of species managed using these methods include lobsters and bivalve molluscs in the Atlantic, mackerel in the Gulf of Mexico and Atlantic Ocean, and sablefish in the Pacific Ocean.

Age-Structured Models (39%)—These methods are used for 8 of the top 10 (by weight) U.S. commercial fish species. The models include ADAPT (24% of category), Cansar (1%), CAGEAN (13%), Stock Synthesis (28%), and virtual population analysis (33%).

Stock Assessment Methods Used by States and Interstate Commissions

Interstate fishery management commissions were created to manage stocks shared among states within their coastal waters (including 3 miles from shore on open coasts, bays, and estuaries). Stock assessments are conducted for species in Atlantic state waters by the Atlantic States Marine Fisheries Commission (ASMFC). ASMFC manages 18 Atlantic coast species, with cooperation among states, NMFS, the U.S. Fish and Wildlife Service, the District of Columbia, and the Potomac River Fisheries Commission. ASMFC supports stock assessment committees for individual stocks, whose work is reviewed by a technical committee for each managed species. The technical committee uses information from the stock assessment to develop options for management measures. Species management boards select and implement a specific management strategy for the species. ASMFC has provided a list of its concerns about stock assessment procedures, which appears in Appendix H.

The Pacific States Marine Fisheries Commission (PSMFC) does not conduct assessments; instead, assessments are conducted by the individual states or by NMFS. PSMFC's primary functions are to maintain fishery databases that contribute to Pacific Coast fishery assessments and to coordinate the fishery management activities of California, Oregon, Washington, Idaho, and Alaska, including interjurisdictional planning among its member states for nonfederal fisheries. Oregon conducts assessments for pink shrimp, red sea urchins, Dungeness crabs, and Pacific herring (Table H.2). Alaska supports a large stock assessment effort for marine organisms harvested from state waters. Tables H.3, H.4, and H.5 show the methods used for assessment of Alaskan shellfish, Pacific salmon, and Pacific herring. Concerns of the PSMFC and of Alaskan stock assessment scientists are given in Appendix H. Finally, the Gulf States Marine Fisheries Commission (GSMFC) operates similarly to the PSMFC; states in this region (e.g., Florida) conduct their own stock assessments and manage a number of species in their own waters. Concerns of the GSMFC also can be found in Appendix H.

Evaluation of Alternative Actions and Specification of Performance Indicators

After data have been collected and the size and population structure of stocks have been estimated, harvest and management strategies must be developed (Chapter 4). The fishery management task is complicated by the fact that many fish stocks overlap the state-federal jurisdictional boundary at 3 miles from shore and state-to-state boundaries. For stocks that are shared by states and the federal government and feature a combination of commercial and recreational fishers, there are sometimes joint management activities and data sharing to allow management of fish populations as complete units.

Tools used by managers of commercial activity vary from fishery to fishery, ranging from no control to specifying total allowable catch that can be removed from a fish population each year. Management measures can also include the amount of fishing and times of the year that fishing can take place; the gear that can be used;

minimum size limits; and in some cases, the amount of fish that any single fisher, community, company, or other entity can catch. Recreational fisheries more often impose minimum size limits, daily catch limits, seasons, and sometimes gear restrictions and requirements to release fish that are caught.

Presentation of Results

A variety of approaches are used to present stock assessment results to managers. For fisheries managed by the regional fishery management councils, NMFS is responsible for preparing Stock Assessment and Fishery Evaluation (SAFE) documents. These documents receive scientific user-group review of varying intensity in different parts of the country. Results are presented in annual or periodic stock assessments. Stock assessment results are summarized in various NMFS publications, such as *Our Living Oceans* (NMFS, 1996). In many cases, there are also public presentations of stock assessment results. Ultimately, stock assessment results are presented to the regional fishery management councils or the boards that govern other fishery management organizations. Stock assessment documents generally contain information about a fishery, its catch and CPUE, survey biomass estimates, and estimates of other important population parameters. When sufficient information is available, population models (see Chapter 3) are constructed to make better use of the information. Frequently, recommendations about acceptable harvest levels are deduced from the population models or "yield or spawning biomass per recruit" studies. Although not yet common, risk assessment and explicit treatment of uncertainty is emerging as an important component of the reports.

APPROACH OF THE COMMITTEE

In 1994, the National Oceanic and Atmospheric Administration (NOAA) asked the NRC to evaluate the scientific basis of U.S. management of Atlantic bluefin tuna and to recommend research to resolve remaining stock structure issues. The NRC completed its work in time for the November 1994 meeting of the International Commission for the Conservation of Atlantic Tunas (ICCAT) and published a report entitled *An Assessment of Atlantic Bluefin Tuna* (NRC, 1994b). The central conclusion of the report was that there was no basis for the use of a stock assessment model that assumed relatively unmixed eastern and western Atlantic stocks of tuna, which was then the practice. The report suggested that NMFS test different models for analyzing tuna data (e.g., Stock Synthesis) and described several technical issues involved in the proper conduct of bluefin tuna assessments, including choice of transformation constants, error structures, weighting, and techniques to standardize catch rates.

As a result of the impact of the 1994 report, the NRC began to receive requests to review stock assessments and models used in other U.S. fisheries. In discussions with NOAA, it was determined that the NRC could play a more strategic role by conducting a scientific review of all stock assessment methods and models used for U.S. marine fishery management, rather than examining stock assessments one by one (see Appendix A for letter of request). The NRC Committee on Fish Stock Assessment Methods was formed in early 1996 to conduct the review (see Appendix B for committee biographies). The committee chose to evaluate existing approaches by asking stock assessment analysts to conduct blind runs of simulated data sets through existing models, an approach similar to that used by the International Council for Exploration of the Sea (ICES) to evaluate its stock assessment models (ICES, 1993). The features of the simulated data sets were known only partially to the analysts and were designed to include characteristics that would test the sensitivity and robustness of the models to realistic permutations of data quality and type.

The remainder of this report elaborates how data are obtained for stock assessments (Chapter 2); how assessments are conducted, including various assessment issues (Chapter 3); and how harvest strategies are developed based on assessments (Chapter 4). The procedure for the committee's simulation is discussed in detail in Chapter 5 and Appendix E. Results of the simulations are also presented in Chapter 5. The committee used the results of the simulations and its knowledge of existing U.S. stock assessment practices to identify new approaches to stock assessments and to recommend other ways in which the assessment of fish stocks could be improved in the United States (Chapter 6). Stock assessment methods are relatively uniform worldwide, so the information in this report may be useful for stock assessment scientists in other nations.

2

Data

Fishery management is based on estimation of either absolute or relative number, biomass, and productivity of fish of a target species available for harvest. Management of fish stocks in a sustainable manner also requires characterization of the population structure of the target species.

Most fisheries include large numbers of commercial fishers and/or recreational anglers. It is almost never possible to monitor the landings of every participant in a fishery or to count and measure every fish in a stock using nonlethal means; therefore, some form of survey sampling is required to collect the data necessary for management. Sample surveys of commercial and recreational fisheries and fishery-independent data are used routinely to characterize numbers of fish and fish population characteristics (see Table 2.1 for a summary of survey characteristics). Surveys are vital for fisheries management; they must be planned with appropriate statistical design and executed with appropriate vessels to gather survey data that are comparable over time (see ASMFC, 1997). Proper data collection requires significant resources (i.e., for sampling, data processing and editing, and database design and management), and these resources are often imperiled when agency budgets must be reduced, leading to degradation in the usefulness of the data being collected. Data collection and processing activities must be protected because the entire fishery management enterprise rests on a foundation of sound scientific data used in appropriate models to characterize the status of exploited populations accurately. The chapter concludes with a short discussion of the importance of considering environmental factors in sampling design. Although this chapter treats commercial and recreational assessments separately, it is imperative that assessment scientists remember that two major types of fishers harvest from shared stocks. Surveys and analysis for recreational and commercial fisheries should be conducted in a compatible manner so the data can be used together. The committee did not evaluate recent efforts to improve fishery-dependent data (e.g., the Atlantic Coastal Cooperative Statistics Program), but believes that fishery-dependent data that do not suffer from the shortcomings tested in this report could improve stock assessments substantially.

BASICS OF SAMPLE SURVEY DESIGN

Survey sampling techniques almost always employ a probability- or design-based sampling scheme, the basic ingredients of which are as follows (Cochran, 1977; Thompson, 1992):

- A finite population of unique and identifiable sample units (e.g., fishing vessels, area of bottom swept by

TABLE 2.1 Properties of Design-Based Sampling Approaches to Estimate Catches, CPUE, and Biological Characteristics from Fishery-Dependent and Independent Sampling

Property	Fishery Dependent		Fishery Independent
	Commercial	Recreational	Trawl or Hydroacoustic Surveys
Sampling frames[a] available	1. List of licensed fishers 2. Spatiotemporal frame of ports and season	1. List of licensed anglers 2. Spatiotemporal frame of access points, fishing areas, and season	Spatiotemporal frame of fish habitat
Sampling units	1. Boat captains 2. Ports-boats and season	1. Anglers or households 2. Access points or fishery areas and season	1. Swept areas and season 2. Acoustic transects and season
Proportion of sampling units actually sampled (n/N)	High	Low	Very Low
Properties of spatial sampling points	Few and well-defined	Many, defined, and diffuse	Countless: large areas in 3 dimensions
Biases	1. List frames Misspecification Underreporting Refusals Misrepresentation 2. Spatiotemporal frames Misspecification Refusals Port avoidance Misrepresentation	1. List frames Misspecification Prestige inflation Nonresponse Refusals 2. Spatiotemporal frames Avidity oversampling Frame undercoverage Refusals	1. List frames: not applicable 2. Spatiotemporal frames Fish migration and aggregation Gear selectivity Insufficient coverage
Precision	High to medium	Medium to low	Low
Potential for data degradation from increased regulation	High: requires a great deal of fisher cooperation	High to medium: requires some fisher cooperation	Low: all sampling under complete control of samplers
Calculation of catch	Direct expansion Ratio expansion	Direct expansion Ratio expansion	Area expansion (No. of fish per area x area)

NOTE: CPUE = catch per unit effort.

[a]The list of all possible sample units is referred to as the *sampling frame*. Sampling frames can be divided into two broad categories: (1) *list frames* and (2) *spatiotemporal frames*.

standard trawl gear)

    • For each unit of this population, a characteristic or group of characteristics (e.g., numbers of fish of a specific species, size composition) that can be observed without error (see Cochran, 1977 on measurement error in sample surveys)

        • A sampling plan that assigns a known probability of selection to each of the sample units

    The list of all possible sample units in the target population is referred to as the *sampling frame*. For example, in sampling commercial fishing vessels at any specific time, the sample units are individual vessels and the sampling frame is the list of all active commercial fishing vessels at that time. Sampling frames can be divided into two broad categories: (1) *list frames* and (2) *spatiotemporal frames*. In the example above, the vessel owner would be selected from a list of owner names, addresses, and telephone numbers, constituting a list frame. If no reliable list frame of vessels and their owners exists, ports could be sampled randomly on days that vessels land their catches, an example of a spatiotemporal frame.

    Fisheries textbooks have recognized the sampling issues involved in stock assessment for some time (e.g., Gulland, 1966; 1969); more recent texts have further highlighted the importance of formal statistical sampling and its terminology (see Gunderson, 1993 for research surveys, Pollock et al., 1994 for recreational surveys, and Fabrizio and Richards, 1996 for commercial surveys). Even today no uniform terminology exists for methods that use the two kinds of sampling frame. Methods that rely on list frames are called *indirect* methods for commercial data collection and *off-site* methods for recreational data collection. In both cases, interviews of anglers or vessel captains are conducted away from the landing site, well after the fishing trip has been completed and without direct observation of the landings by the survey agent. Methods that rely on spatiotemporal frames are called *direct* methods for commercial data collection and *on-site* methods for recreational data collection because interviews of anglers or vessel captains are conducted as fish are landed and the landed fish can be observed directly, measured, and weighed. Time-tested approaches based on the sampling design selected are used to sample from both types of sampling frames.

    The basic survey design is the simple random sample (SRS) design that assigns an equal probability of selection to each sample unit in the sampling frame. However, more complex designs are usually required for applications where landings or anglers must be sampled by geographic area within specific time periods (a spatiotemporal frame). Whatever design is used, the sampling plan or design assigns probabilities to the observations so that a probability distribution such as the normal or Poisson need not be assumed.

    Properties of the estimates, such as bias or precision, are evaluated strictly as functions of the sampling design. Such properties are referred to as *design-based* to distinguish them from *model-based* properties that depend on a statistical distribution for their validity. The estimate of the mean from a stratified random design is more precise than the mean from an SRS design if the variance *within* strata is less than that *between* strata and samples have been allocated to strata in an optimal fashion. In a model-based design, the estimate of the mean defined for the Poisson distribution is the most precise estimate of the mean only when the observations exhibit a Poisson distribution.

    The degree of precision for one design relative to another is referred to as the design efficiency. There are two ways of increasing the precision of the mean (i.e., decreasing the standard error). The first way would be to increase the overall sample size; the second involves keeping the same sample size, but changing the number of samples allocated to each stratum. The following sections discuss commercial surveys, recreational surveys, and fishery-independent data, including the purpose for using each method, the types of data collected, the sampling methods, and the limitations of the method.

COMMERCIAL SURVEYS

    The worldwide decline in many fish stocks is reflected in U.S. fisheries. The status of 275 stock groups reported by the National Marine Fisheries Service (NMFS, 1996) reveals that of the 181 that have a known long-term potential yield (LTPY), 55% are exploited near or above the sustainable level. Overexploitation has resulted in the collapse and/or restriction of important U.S. fisheries and has had a negative impact on the quality of data obtained from commercial catches because fishing effort is constricted. Shorter seasons and diminished fleets can directly result in degraded data. Very short seasons yield the narrowest views of annual changes in abundance and availability in addition to diminished sampling opportunities. Likewise, diminished fleets may change how fish are targeted, and the fishing power of the fleet and effort patterns may change; such changes can be difficult to document.

Degradation in data quality also results when fishers do not trust stock assessments or other fishery management because they believe that such activities are responsible for increasingly restricted quotas, shortened seasons, and diminished opportunities. When stocks are overexploited, catch and effort quotas are mandated through fishery management plans to allow the stocks to increase. Hence, even fewer fish are available for harvest during times when the stocks are already at low levels, which reduces the chance that fishers can maintain their livelihoods. Typically, restrictive quotas result in an adversarial relationship between the fisher and the management agency and often results in refusals to cooperate in data collection except when mandated by law. Even with legal requirements, data that are reported cannot necessarily be assumed to be accurate.

Purpose

All age-structured assessment models require reasonably accurate estimates of the number, weights, and ages of fish removed from the population. Removals include fish that are landed as well as those that are killed through contact with fishing gear and those that are caught and discarded. Models must also account for landings that have been misreported with respect to area and species.

Data Collected

Basic data are collected from commercial fishers regarding vessel name and characteristics; gear; location(s) and date(s) fished; fishing effort; time, weight, and length of target species landed; weight of sample; and otoliths or scales from sampled fish.[*] Total weight of a species landed by commercial fishers is usually estimated from purchase slips (*fish tickets*) or logbooks carried by fishing vessels. On-board observers representing the management agency (the National Marine Fisheries Service in the United States) provide information about the amount of fish caught in some fisheries (e.g., Northeast sinknet), including discards. The incidence of highgrading[†] and misreporting can be inferred from comparisons of the observed catch and logbook entries. Size and age composition of the catch are usually determined by on-board observers or by agency representatives who sample the catch at the point of landing.

Sampling Methods

Off-Site Methods

The most inexpensive methods of obtaining commercial fisheries data are from reports of fishers about their landings. These methods include purchase slips, logbooks, and trip tickets. Such self-reporting methods are relied on heavily in commercial fisheries in the United States. The collection of purchase slips and logbooks is a complete census if fishery regulation requires submission of these by all fishers as a condition of fishing and as such does not require sampling of the list of licensed fishermen. However, it is prudent to estimate potential bias in these self-reported data by using on-site sampling to validate reported landings.

In the U.S. Northeast region, the collection of fishery landings and effort data was changed from a voluntary to a mandatory program in 1994 for many of the species regulated under federal fishery management programs.

---

[*]See http://remora.ssp.nmfs.gov for commercial fisheries data for U.S. Atlantic and Gulf coasts. Otoliths are ear bones that can be analyzed to estimate the ages of individual fish.

[†]Highgrading is the practice of discarding less valuable fish as fishing operations proceed in order to achieve the highest value catch for a given number or weight of fish.

Prior to 1994, these data were collected from participating fish dealers and vessels through a network of port agents and biological samplers (Burns et al., 1983). After 1994, dealer data were collected by or submitted to the port agents and logbooks were submitted by all permitted vessels. These data are augmented by those collected by many state agencies for their inshore fisheries. Requirements for sampling of landings reflect season, gear type, and area. Case studies are reported in Burns et al. (1983), Quinn et al. (1983), and Crone (1995).

On-Site Methods

Even though on-site (or direct) methods are more expensive than off-site methods, they are the only way to obtain reliable data on length and age or to validate fishers' self-reported catches. The two methods used in on-site sampling of commercial fisheries are port sampling (Burns et al., 1983) and observers on-board fishing vessels (Murawski et al., 1994).

Biological sampling of size, age, and sexual maturity has been conducted at the larger ports for more than 100 years (Murawski et al., 1994). For the Northeast region this included 21 ports (Burns et al., 1983). Samples to determine length and age are generally obtained in a two-phase scheme. In the first phase, a large simple random sample is selected, and each fish is assigned to a *stratum* according to its length. In the second phase, a much smaller simple random sample is chosen from each stratum of similar-sized fish. Otoliths or scales are removed from these second-phase fish to determine their ages. The International Convention of the Northwest Atlantic Fisheries recommended a minimum of one age sample per 1,000 tons of landings for a typical species.

There is some disagreement about the comparability of age and length observations from observer versus shore-based operations. Baird and Stevenson (1983) concluded that the precision of the estimates of the numbers-at-length did not differ significantly between the two sampling approaches. However, Zwanenburg and Smith (1983) showed that actual estimates of the numbers-at-length, where both shore-based and observer samples were available, differed significantly. These differences were not consistent with respect to size class as might be expected if highgrading or discarding of certain classes had occurred. Instead, the differences were interpreted to reflect either spatial heterogeneity of size classes in the ocean or storage-induced variation due to onboard processing methods such as gutting and freezing.

The most direct method of obtaining data for the entire catch is by deploying observers on-board fishing vessels. On-board observers can record species composition, weights of discards and landings, and areas and time fished. Additionally, on-board observers can determine length and gender of fish and can collect otolith samples to be used for catch-at-age and ageing indices. An indirect benefit of on-board observers is that interactions of fishing gear and protected species can be observed, as well as any potential violations of conservation measures. It is less likely that highgrading or misreporting of catch, bycatch, or other data in logbooks will occur if an observer is on-board during fishing operations. Although this method provides the most reliable data, it is the most expensive and requires relatively well-trained personnel to manage and report the data accurately. Observer programs are conducted for some Atlantic and Pacific Ocean fisheries on both domestic and foreign vessels. The costs of observer programs are borne either by the federal government or by the fishers themselves. Kulka and Waldron (1983) discuss an observer program as it was originally envisaged in Atlantic Canada in the early 1980s. Although no specific information is given by Murawski et al. (1994) on how vessels were selected to use observers, one method would be to stratify by vessel size and gear type. Within each stratum, vessels sampled are chosen in some statistically appropriate manner. The success of observer programs depends on establishing a statistically valid method for sampling vessels, adequate training of observers, appropriate reporting procedures, and the cooperation of vessel owners and crew during sampling procedures.

Limitations

Sampling programs and designs for commercial fisheries have not been analyzed extensively in the scientific literature, apart from the case studies listed above and some of the references given below. Stanley (1992) used a bootstrap technique to characterize the variation associated with catch per unit effort (CPUE) from logbooks.

Bayesian methods for optimizing two-phase sampling schemes are discussed in Smith and Sedransk (1982), Jinn et al. (1987), and Nandram et al. (1995).

Sen (1985) used regression techniques to demonstrate that two-phase sampling offers little advantage over simple random sampling for estimating age composition. Smith (1989) discusses conditions for which simple random sampling conducted in a single phase gives estimates of age composition that are as accurate as the estimates obtained from two-phase sampling, based on the relative per-unit costs of obtaining the stratifying variable (length or weight in his case) and the otolith. The impacts of sampling variation on stock status measures are evaluated in Pelletier and Gros (1991) and Nandram et al. (1997).

RECREATIONAL SURVEYS

Although catches in marine recreational fisheries have increased in the United States since the 1950s, the probability of the continuing growth of this sector is unknown. Coastal populations in the United States are expected to increase about 13% between 1990 and 2010 (NOAA, 1990), which will result in increased growth in use of natural resources and increased negative impacts on coastal habitats and water quality. However, the effect of population increases on recreational fishing may be diminished by the decline in fishing activity as the population ages. Levels of recreational angling in the future will also depend on anglers' perceptions of stock status.

Purpose

Surveys of anglers have been used in the past to estimate total catch, effort, and CPUE to support management. These data have been used empirically to characterize the fishery, in combination with abundance estimates obtained from fishery-independent surveys. Angler catch data are also used to assess compliance with regulations. Before implementation of the Marine Recreational Fishery Statistics Survey (MRFSS)[*] by the National Marine Fisheries Service in 1979, recreational data were rarely used to produce quantitative projections of recruitment, fishing mortality, and exploitation rates of recreational fisheries.

Ideally, surveys of recreational fisheries should provide consistent annual estimates of total catch and effort of the larger recreational fisheries. Surveys should use standardized sampling methods so that total catch, yield, and effort data are comparable from one year to the next. Spatial and temporal frames should be the same as for commercial fisheries so that data can be combined for assessments of stocks that are used by both commercial and recreational fishers.

Although catch-at-age and surplus production models have been the mainstay of commercial fishery management, they are only beginning to be used in recreational fisheries (e.g., Quinn and Szarzi, 1993). The main difficulty has been that it is very expensive to obtain appropriate data from recreational fisheries to support catch-at-age modeling approaches. Additionally, both types of models require consecutive years of fishery monitoring, and aside from MRFSS, such long-term monitoring surveys are rare for recreational fisheries (Fry, 1949; Serns, 1986; Carl et al., 1991). Sampling must be sufficiently intensive to estimate short-season commercial (including migratory stocks) and recreational fisheries. Even MRFSS has been unable to sample these sectors of recreational fisheries adequately. Scientists concerned with recreational fisheries (especially in South Africa and New Zealand) have tended to base their recommendations on yield-per-recruit and spawner biomass-per-recruit analyses.

Recreational fisheries differ fundamentally from commercial fisheries in geographic extent and in the skill and motivation of fishers. Anglers typically access the water from many more launch sites than do commercial fishers. Commercial vessels are typically larger and their operations require greater infrastructural support; thus, commercial vessels concentrate in larger, better developed ports. In contrast, anglers have smaller vessels and can launch from smaller, more diffuse access points. Recreational fishing also takes place from public and private piers, bulkheads, and shorelines. Respective skill levels of the two types of fishers are very different. Commercial fishers must have sufficient skill in fish harvesting to offset their operational costs, whereas anglers need not offset

---

[*]See http://remora.ssp.nmfs.gov/mrfss/index.html for additional information and data.

any costs. A day on the water is rewarding to many anglers even when no fish are landed. Hence, in recreational fishing there are no dire personal economic consequence for inefficient harvesting of fish.

Data Collected

The types of data generally collected in surveys of recreational fisheries are summarized in Pollock et al. (1994). These data include target species, number of fish caught, and number of angler trips. Collecting data on fish length, weight, and age is more difficult for recreational fisheries than for commercial fisheries; furthermore, these data can be collected only by using expensive access point surveys. Data that result in fish mutilation, such as removing the otoliths for age determination or opening the body cavity to determine reproductive maturity, are more difficult to obtain from anglers who may value the trophy qualities of their catch. Finally, input data to fisheries models rely on long-term collection of catch and effort data. Prior to MRFSS, such time series were rare for recreational fisheries.

Sampling Methods

Sampling approaches in recreational fisheries vary in terms of the location and timing of the angler interview. Interviews take place either on-site or off-site. The choice of interview location is determined by the available budget and the quality and quantity of data needed to meet survey and modeling objectives. The sampling frame and the definition and method of selection of sampling units are determined by whether interviews are conducted on- or off-site. The choice of on- or off-site interviews controls the type of data biases that may be experienced.

On-Site Methods

On-site interviews provide more reliable data than off-site surveys (Pollock et al., 1994). When the angler is interviewed on-site, the survey agent can actually examine the fish, take biological measurements, and record the number of angling trips. Interviews can take place either at an access site upon completion of the angler's trip or during fishing by a roving survey agent. In either case, these survey methods are relatively expensive; only trained interviewers can be used and their travel can be costly.

On-site methods rely on spatiotemporal sampling frames; hence, the sampling frame is delineated by the geographic extent of the fishery and the timing and length of the fishing season. Once the sampling frame is defined, sampling units are determined as day-place combinations and can be chosen randomly following standard sampling procedures outlined in Cochran (1977) and Thompson (1992). Selection of sampling units (day-place locations) for on-site surveys is typically done with stratification procedures to correct for spatial and temporal patterns in recreational fisheries (Malvestuto et al., 1978; Malvestuto, 1983; Pollock et al., 1994). Anglers are more likely to fish on their days off from work; therefore, weekend days have more fishing trips than weekdays. Similarly, the opening week of the season in some fisheries has heavier effort than the weeks following. In boat-based recreational fisheries in which trips last many hours, more trips will be completed in the afternoons than in the mornings. To provide more precise variance estimation, survey sampling effort is matched with sampling units that have more trips (Jones and Robson, 1991). Therefore, weekends, opening weeks, and afternoons are usually selected more frequently by using nonuniform selection probabilities. Recent advances have improved sampling designs for on-site surveys (Robson and Jones, 1989; Jones et al., 1990; Hayne, 1991; Robson, 1991; Wade et al., 1991; Hoenig et al., 1993).

Off-Site Methods

When budget considerations outweigh the need for accurate catch data, off-site methods are preferred, especially if ancillary economic data are desired (Brown, 1991; Essig and Holliday, 1991; Pollock et al., 1994). Off-site surveys take place after the fishing trip has been completed and the angler has returned home. Methods include mail, telephone, door-to-door (rarely used), diary, and logbook surveys. Except for door-to-door surveys, these methods are inexpensive. Sampling unit selection in off-site surveys is straightforward. Typically, anglers are chosen for telephone surveys from list frames by systematic sampling and simple random sampling by random-digit dialing. These methods are described in Dillman (1978), Waksberg (1978), and Essig and Holliday (1991). Survey methods may also rely on lists of angler names, addresses, and telephone numbers from fishing licenses. The sampled population is defined as individuals participating in recreational fishing. Sampling units are anglers or angling households. Selection of sampling units from list frames or random-digit dialing is well understood and documented (Dillman, 1978; Waksberg, 1978; Frey, 1983; Groves et al., 1988; Lepkowski, 1988; Weithman, 1991).

Regardless of whether an on- or an off-site survey method is chosen, biases will be present and must be corrected to produce accurate estimates of catch and effort. In on-site surveys, two types of bias occur: avidity and length of stay. On-site surveys encounter avid anglers more frequently than casual anglers and this can result in biases of the angling population's demographics and economics. Similarly, during a roving on-site survey, anglers with longer trips will be sampled disproportionately; if their catch rates are not representative of all trips, estimates will be biased. The primary biases of off-site surveys are their reliance on anglers' self-reporting and quickly outdated list frames (Groves, 1989). Inherent in self-reporting is the tendency for anglers to exaggerate their catches, forget trips, and misidentify species; hence, data quality degrades over time, especially for catch data. A combination of on- and off-site survey methods is often used to minimize costs and inherent biases and to maximize data quality (Pollock et al., 1994). For example, MRFSS uses a telephone survey with random-digit dialing to obtain effort data and an access point survey to obtain catch rate data (Essig and Holliday, 1991). To produce estimates of total catch, catch rate is multiplied by effort. Such combined surveys are often the most efficient for marine recreational fisheries.

Limitations

Aside from biases inherent in recreational data mentioned previously, violation of the assumptions underlying the relationship between catch and population size can compromise the accuracy of model predictions. Many stock assessment models assume that

$$C = qfN \tag{2.1}$$

where C is catch, q is the catchability coefficient, f is fishing effort, and N is average stock abundance. Sometimes q is assumed to be constant. In order for CPUE data to be used as an accurate index of abundance, it is necessary to conduct standardization studies. Because scientific surveys can be designed to collect data in a consistent manner, they will usually provide more consistent information than commercial or recreational CPUE. But this need not always be the case (e.g., Quinn, 1985). Although much research has been conducted regarding the catchability coefficient in marine commercial fisheries (Bowman and Bowman, 1980; Winters and Wheeler, 1985; Crecco and Overholtz, 1990; Gordoa and Hightower, 1991; Rose and Legget, 1991; Swain and Sinclair, 1994; Hannah, 1995) and freshwater recreational and commercial fisheries (Inman et al., 1977; Peterman and Steer, 1981; Brauhn and Kincaid, 1982; Nielsen, 1983; Crecco and Savoy, 1985; Engstrom-Heg, 1986; Collins, 1987; Kleinsasser et al., 1990; Borgstrom, 1992; Buijse et al., 1992; Shardlow, 1993), only a few studies have focused on marine recreational fisheries (Loesch et al., 1982; Claytor et al., 1991; Pickett and Pawson, 1991; Kerr, 1992).

The assumptions that q can be estimated accurately or is constant and that C is linear with respect to N are often likely to be untrue for marine recreational fisheries. For example, the increased construction of artificial reefs may make fish more vulnerable to capture (Kerr, 1992), thereby increasing q. The catchability coefficient will be influenced by the diversity of fishing practices (Pickett and Pawson, 1991), resulting in estimates of q with large standard errors and, potentially, nonlinearity in the relationship between C and N. Additionally, freshwater studies have shown that different strains of fish stocked have different q values (Dwyer, 1990; Pawson, 1991). Similarly, stock differences in marine fishes may also result in differences in q. Finally, angler misreporting of catch rates will influence the accuracy of estimates of q. In an Alaskan razor clam study, CPUE did not correlate well with survey abundance because the harvest by each clam digger tended to depend on the skill and fitness of the individual, irrespective of the density of clams (Szarzi et al., 1995).

Many stock assessment models use CPUE as a relative measure of population abundance. In commercial fisheries, with their relatively limited range of fishing methods, effort data can be standardized to reflect a common effort unit. Standardization factors can be developed through field experiments designed to measure differences in the relationship between effort and abundance for different types of fishing gear, which results in standardization factors. General linear models are used frequently to compare different gear types, which likewise results in standardization of fishing effort (Hilborn and Walters, 1992; Quinn and Deriso, in press).

In recreational fisheries, standardization of fishing effort is a major problem because of the different gear used, angler skill, fish behavior, and logistic problems. Estimation procedures for recreational CPUE are poorly understood (Jones et al., 1995). Fortunately, MRFSS has relied on access points to obtain CPUE data (equal probability) and used the *ratio-of-means* estimator (the correct estimator within a stratum), which could be a single day's survey of a specific region. Estimates of catch and effort in recreational fisheries are characteristically more imprecise than those of commercial fisheries. Two characteristics of recreational fishing underlie this imprecision: (1) access to the fishery typically occurs through a greater number of locations and it requires more sampling to characterize, and (2) skill levels of anglers vary greatly and are intrinsically more variable. Sampling surveys can be designed to maximize precision when increased precision is a recognized objective of the survey.

The principal designed survey of U.S. marine recreational fisheries, MRFSS, uses a complemented design to obtain estimates of catch and effort. The complemented design consists of telephone interviews of coastal-county residents to estimate effort, and on-site access-point interviews to estimate catch rates and non-coastal resident participation. Because telephone interviews are relatively inexpensive, precise estimates of effort can be obtained at low cost. In contrast, on-site interviews are relatively costly to obtain and smaller sample sizes can be obtained; therefore, catch rate estimates are less precise. The objectives of MRFSS were to provide estimates of overall recreational catch and effort with broad scale precision. The MRFSS design does not provide precise estimates for the types of angling that require a specialized and targeted survey such as fishing on highly migratory species, charter-boat fisheries, or for species available only during a short fishing season. Getting interviews in these fisheries becomes more unpredictable and results in estimates with high uncertainty.

In most mixed-use fisheries, commercial catch dominates total catch. For stock assessments in such mixed-use fisheries, imprecision in recreational catch estimates add little to the overall uncertainty in total catch. However, in a few fisheries recreational catch is a considerable or predominant component of total catch, for example, for bluefish and striped bass. Such fisheries cannot be surveyed with sufficient precision with the generalized survey approach of MRFSS and a specially designed survey could result in increased precision of catch and effort estimates if the additional cost is deemed justified. Even beyond these concerns, variance estimates in ratio estimators are compromised when catch and effort are correlated (Cochran, 1977; Jones et al., 1995). This is also true for commercial CPUE data. In summary, the reliability of CPUE estimates from recreational fisheries must be understood better before these data are used extensively in stock assessment models.

FISHERY-INDEPENDENT DATA

Purpose

The major objective of fishery-independent surveys is to monitor temporal and spatial changes in the relative or absolute abundance of a target fish population or a particular component of that population (e.g., larvae, juvenile, spawning adults) in a manner that is not subject to the biases inherent in commercial or recreational fishery data. Gunderson (1993) presents an overview of the major types of fishery-independent surveys being conducted at present, which include surveys of juveniles and adults by trawl, acoustic, aerial, and SCUBA-based methods and surveys of eggs and larvae by plankton tows (see also Helser and Hayes, 1995).

For any component of the fish population being targeted, the ideal survey should maintain the same gear, area of coverage, and time period throughout the time series of the survey, provided the survey covers the entire geographic range of the population. It is generally assumed that keeping all of these operational procedures constant over time allows the interpretation that year-to-year changes in measured abundance indicate true or relative changes in population size. The area of coverage and time period must be defined such that the targeted component of the population is consistently available to the survey. When a survey must be changed, calibration experiments have to be performed to make old and new survey methods comparable and to maintain the continuity of the data series.

Data Collected

The types of data generally collected on fishery-independent trawl and acoustic surveys are reviewed in Gunderson (1993). These data include (for each target species) numbers and weight of fish caught; length and age compositions; and biological information such as gender, maturity, fecundity, and condition. Most trawl surveys conducted for stock assessment purposes also collect abundance and demographic information for other species captured and therefore offer an opportunity for ecosystem-based assessments over broad temporal and spatial scales. In addition, a number of hydrographic variables may be measured routinely, such as depth, surface and bottom temperature and salinity, oxygen, and the concentration of various nutrients. Full-depth profiles of these characteristics may be measured. Such environmental data are becoming more important as greater effort is devoted to relating fish catch to concurrently measured hydrographic variables as a means of explaining spatial distribution and varying availability and catchability as a function of environmental variables (e.g., Murawski and Finn, 1988; Sinclair, 1992; D'Amours, 1993; Perry and Smith, 1994; Swain and Kramer, 1995). Smith et al. (1991) and Smith and Page (1996) used environmental relationships to explain trends in either the survey catch or abundance of Atlantic cod (*Gadus morhua*) as estimated by a survey.

Sampling Methods

Approaches used to sample and estimate abundance for fishery-independent monitoring programs are of two general types. In the first type, survey sampling methods dictate the survey design. Station locations (whether for trawl sets, dredge sets, or acoustic transects) are chosen randomly from a sampling frame of possible locations in the study area. Examples of surveys in this category are the trawl surveys of groundfish for the eastern coasts of Canada (Doubleday and Rivald, 1981) and the United States (Azarovitz, 1981), dredge surveys of scallops on Georges Bank (Mohn et al., 1987) and the northeastern United States (Serchuk and Wigley, 1986), and acoustic surveys of pelagic fish off South Africa (Jolly and Hampton, 1990).

Stratified random designs are the most common designs used for fishery-independent surveys (Gunderson, 1993). Strata are usually defined by water depth or species' distributional patterns but also may reflect management boundaries. Smith and Gavaris (1993a) found that even when strata had been designed to correspond to distributional information for Atlantic cod, most of the gain in precision was obtained from how samples were allocated to strata and not from increased homogeneity within strata.

As noted earlier, the optimal allocation of samples to strata is in proportion to the expected variance in each stratum. Smith and Gavaris (1993a) report on the success of using results from previous years' surveys to design an optimal allocation scheme for the current year for a survey of cod. In addition, adaptive *allocation* methods have been developed whereby in the same survey, a portion of the total number of samples is used to characterize the variance within strata, followed by a secondary allocation of the remaining samples to more variable strata (Francis, 1984; Jolly and Hampton, 1990).

Conversely, adaptive *sampling* attempts to increase precision by conditioning the selection probabilities of samples on the observed values (Thompson and Seber, 1996). A condition of interest is defined such that sampling continues in an area until a second condition defined as a *stopping rule* has been reached. For example, the condition of interest could be defined to be "catches greater than 100 kg." Once a catch exceeding this value has been obtained, samples are collected in adjacent areas until *n* samples have been collected, two or more catches of less than 100 kg have been observed, or some other stopping rule has been reached. Thompson (1992) presents the estimation theory required to accommodate the above scheme within design-based theory. Although a number of techniques exist for increasing the efficiency of stratified surveys, most long-term surveys of this type do not change their strata or allocation design over time (an exception is reported in Smith and Gavaris, 1993a). One major limitation of modifying the survey design to increase its efficiency is that most surveys are multispecies in focus. Changes in the design to improve sampling for one species may not improve the design with respect to the precision of estimates for other species.

Some scientists believe that spatial methods are preferable for analyzing survey data because design-based methods assume spatial independence for their estimates, especially variances (e.g., Sullivan, 1991; Simard et al., 1992; Ecker and Heltshe, 1994). Although finite population methods may ignore fine-scale spatial structure in the population being sampled, it is not necessarily true that spatial independence is required for properties of the estimates to hold (i.e., concerning bias and precision; see Cochran, 1977; Smith and Robert, 1997). However, spatial structure need not be ignored completely in finite population survey designs, because large-scale spatial structure can be incorporated by using stratification. Fine-scale structure can be incorporated by using a predictive model with covariates (see discussion in Smith and Robert, 1997).

The second main type of survey scheme includes the nonrandom-type survey such as fixed-station trawl surveys (e.g., eastern Bering Sea trawl survey, Traynor et al., 1990) or fixed-transect methods common to acoustic surveys (Simmonds et al., 1992) and also used for other types of gear (e.g., longline survey for halibut; Pelletier and Parma, 1994). Estimation methods are not uniquely determined by the survey design in this type of scheme, and models with implicit (e.g., contouring, Delaunay triangles; Robert et al., 1994) or explicit (e.g., kriging*) spatial structures have been used to estimate abundance (e.g., Conan and Wade, 1989; Guillard et al., 1992; Simard et al., 1992). The fixed-station trawl surveys in Alaskan waters (including walleye pollock [*Theragra chalcogramma*], Traynor et al., 1990) and in the Barents Sea (for cod and haddock [*Melanogrammus aeglefinus*], Korsbrekke et al., 1995) are designed with an areal stratification scheme, and although both use fixed stations, they have been analyzed as though the stations were random.

By definition, fixed-station or fixed-transect surveys maintain the same stations or transects each year. Therefore, properties of the estimators are not based on the sample selection scheme per se and improvements in efficiency and precision are not necessarily made through changes to the survey design. Arguments have been made that the mean from a systematic or fixed survey design can be more precise than the mean from a simple random sample (Cochran, 1977; Hilborn and Walters, 1992, pp. 172-173) or a stratified random sample (Simmonds and Fryer, 1996, pp. 39-40), but the necessary conditions are based on the data's exhibiting a specific spatial pattern. Further, given that the precision of the mean of a systematic survey cannot be estimated unless the exact nature of the spatial pattern is known, one never knows whether changes to the operation of the survey can or did improve the precision of the mean.

Estimates of abundance from a spatial model or some other statistical model of the data can be optimized by using the "best" (minimum variance unbiased) estimates for that particular model. These estimates usually do not incorporate the survey design and hence changes in design do not affect their precision apart from any effects due to changes in sample size. The robustness of the models and associated estimators with respect to likely violations of their basic assumptions have to be assessed.

Most assessment models assume that survey catchability and the relative vulnerability and availability of different age classes stay constant over time, so that survey catch rates (either by age or overall) can be used as indices of abundance. In reality, the assumption of constant catchability and availability with age over time may be violated. Changes in availability refer to changes in area occupied by the target population due to changes in abundance (e.g., Crecco and Overholtz, 1990; Swain and Sinclair, 1994), as well as changes in depth distribution due to diel behavior (Walsh, 1991; Engås and Soldal, 1992; Michalsen et al., 1996) or for other reasons (Godø, 1994). Catchability is used to imply some interaction of the fish with the fishing gear or survey process. Differences in survey catchability could result from size-selective catchability (Godø and Sunnanå, 1992; Aglen and Nakken, 1994); environmental effects (He, 1991; Smith and Page, 1996); changes in horizontal and/or vertical distribution due to the noise of the survey vessel or fishing gear (Ona and Godø, 1990); and changes in the configuration of the net with depth (Godø and Engås, 1989; Koeller, 1991). Acoustic surveys may also be affected by some of these factors, with the added problems of changes in target strength (in split- or dual-beam systems) due to changes in the spatial orientation of fish in the acoustic beam (MacLennan and Simmonds, 1992), physiological changes (Ona, 1990), and attenuation of signals close to the seafloor (Ona and Mitson, 1996).

---

*Kriging is a minimum-mean-square-error method of spatial prediction (Cressie, 1993).

Changes in availability due to changes in stock area can be monitored with information from the fishing fleet in the current year. Changes in availability to trawl gear due to diel behavior may be detected by using acoustic and trawl gear concurrently (Godø and Wespestad, 1993). Any significant effects of diel behavior changes on target strength in acoustic surveys require corrections to the data (Traynor and Williamson, 1983).

Methods for correcting trawl data for size-selective catchability are discussed by Godø and Sunnanå (1992), Dickson (1993a,b), and Aglen and Nakken (1994). Corrections due to changes in gear geometry are given in Koeller (1991), among others. Recently, visual observations from submersibles have been used to ascertain catchability for acoustic (Starr et al., 1995) and trawl surveys (Krieger and Sigler, 1996) of rockfish.

Environmental Data

Debates have raged over the magnitude of the effects of fishing versus environmental factors in interpreting survey and fisheries data and in explaining declines in fish populations (e.g., NRC, 1996). In many cases, both factors have probably played a major role. Improved stock assessments may result from more explicit consideration of directional environmental change in sampling strategies and assessment model assumptions.

Global climate can change as a result of natural climate cycles lasting a few years (the El Niño-Southern Oscillation [ENSO]) to millions of years (the Milankovich cycle of Earth orbital variations). Climatic conditions can shift rapidly from one regime to another, which may have occurred in the Pacific Ocean in 1976-1977 (Venrick et al., 1987; Miller et al., 1994; Polovina et al., 1994), or may change more gradually. Climate may also be changing as a result of long-term human action, for example, from the addition of "greenhouse" gases to the atmosphere.

Whatever their causes, climate changes can affect the abundances and distributions of fish stocks (NRC, 1996). Stock assessments must be robust to changes in fish population distributions. During this century, decadal-scale cycles in the North Atlantic Ocean have resulted in alternating warm and cold periods along the European Atlantic Ocean coast. During the cooler decades, gadoid fish species (e.g., cod and haddock) were abundant in the eastern Atlantic Ocean despite heavy fishing pressure. The gadoid fisheries collapsed after 1970, at the time of the return of warmer water and more southern species. Intense fishing pressure may have precipitated the post-1970 collapse, but climate change or changes in Atlantic Ocean circulation patterns could also have contributed to the collapse of these northern species. A similar interdecadal cycle has been observed in the Pacific Ocean. Climate cycles with periods averaging slightly less than 20 years have been observed (Wooster and Hollowed, 1991; Beamish and Bouillon, 1993; Royer, 1993).

Fish abundance shows some relation to weather cycles in California (Soutar and Isaacs, 1974) and in the British Isles (Russell, 1973; Cushing and Dickson, 1976; Southward et al., 1995). Lluch-Belda et al. (1992) documented similar correlations between climatic conditions and sardine and anchovy abundances worldwide. Many of the species observed are not subject to direct fishing mortality, which suggests that the magnitude of climate effects on fish stocks can be as great or greater than the effects of fishing on commercially valuable stocks. Stresses from environmental factors may interact synergistically, resulting in greater natural mortality than would be predicted from the sum of the separate factors. Responding to decreasing survival or shrinking geographic range caused by changes in climate and other environmental factors could require different assessment and management strategies than those now used to detect changes in fish abundance presumed to be caused by overfishing. The accuracy of fish stock assessments can be affected by environmental factors. Assessment methods may under- or overestimate stock size when stock distributions change unless the assessments account for movements of fish stocks as they adjust to changing water temperatures. Survey areas should be large enough to observe changing distributions of the stocks. Decreasing stocks may remain abundant in localized areas. Inadequate sampling of broader areas would then fail to detect the overall stock decreases or would miss remaining high-density areas. Three related assessment issues worthy of continued long-term research follow:

1. What are the effects of climate variability and change on fish stocks? To what extent does climate change drive stock change? The Global Ocean Ecosystems Dynamics program (U.S. GLOBEC, 1995) is largely focused

on this question. Climate change can be slow, and slow changes are particularly difficult for assessment methods to detect and compensate for.

2. As climate prediction improves, can the changes in fish stock abundances and/or distributions due to climate change be predicted?

3. Can these predicted effects be incorporated into fish stock assessments and fishery management strategies?

3

Assessment Methods

REVIEW OF EXISTING METHODS

Assessment methods generally are based on two types of mathematical models: (1) a model of the dynamics of the fish population under consideration, coupled with (2) a model of the relationship of observations to actual attributes of the entire fish population. These models are placed in a statistical framework for estimation of abundance and associated parameters and include assumptions about the kinds of errors that occur in each model and an assumption about the objective function[*] used to choose among alternative parameter values. These errors can be characterized broadly as either process errors or observational errors.

Process errors arise when a deterministic component of a population model inadequately describes population processes. Such errors can be found in the modeled relationships between recruitment (or year-class abundance) and parental spawning biomass and can occur as unpredictable variations in the age-specific fishing mortality rates of an exploited population from year to year. In contrast, observational errors arise in the process of obtaining samples from a fishery or by independent surveys. Decisions must be made regarding the type of probability distribution appropriate for each kind of error (e.g., normal, lognormal, or multinomial) and whether errors are statistically independent, correlated, or autocorrelated.

The form of the objective function chosen for parameter estimation is based on a likelihood function.[†] In the models reviewed in this report, the objective is to maximize the total log-likelihood function. The total log-likelihood is generally a weighted sum of log-likelihood functions corresponding to the different types of observations. It is often simplified to an analogous problem of minimization of weighted least squares.

This report concentrates on complex population models, although the committee acknowledges that less structured stock assessment approaches may be more appropriate for some fisheries, as described in Chapter 1. Examples include the use of linear regression models for some Pacific salmon species and multispecies trend analyses (Saila, 1993).

---

[*]Objective functions measure the "goodness of fit" between the population model and the observations.

[†]A likelihood function gives the probability of obtaining a particular set of data as a function of a model having parameters that are unknown. By maximizing the likelihood with respect to the parameters, one estimates the parameters that provide the highest probability that the data occurred. Similarly, minimizing a sum of squares finds the parameters that make the data and the model predictions as close as possible in terms of squared deviations.

General types of population models include surplus production, delay-difference, age-based, and length-based models (see Chapter 5 for specific references and more detailed descriptions of the models used in this study). They rely on rates of change in biomass and productivity that can be calculated based on information about yield from fisheries, recruitment, and natural deaths. Detailed presentations of these models are given in Ricker (1975), Getz and Haight (1989), Hilborn and Walters (1992), and Quinn and Deriso (in press). Models of population change can be written as a differential equation

$$\dot{B} \equiv \frac{dB}{dt} = \dot{P} - \dot{Y} = \dot{R} + \dot{G} - \dot{D} - \dot{Y} \tag{3.1}$$

or, in words, the rate of change in biomass ($\dot{B}$) equals productivity ($\dot{P}$) minus yield ($\dot{Y}$). The productivity of a population depends on the recruitment of progeny ($\dot{R}$) and the growth ($\dot{G}$) and death ($\dot{D}$) of existing individuals. Barring time-dependent processes in Equation (3.1), equilibrium biomass and yield result only if some of the rates in Equation (3.1) are regulated by population densities; otherwise, the population can either increase or decrease without limit.

Surplus Production Models

This type of model can be implemented with an instantaneous response (no lags) or 1-year difference equation approximation. These models have simple productivity parameters embedded and require no age or length data (Schaefer, 1954; Fletcher, 1978; Prager, 1994). Estimation is accomplished by fitting nonlinear model predictions of exploitable biomass to some indices of exploitable population abundance (usually standardized catch per unit effort, CPUE). The primary advantages of surplus production models are that (1) model parameters can be estimated with simple statistics on aggregate abundance and (2) the models provide a simple response between changes in abundance and changes in productivity. The primary disadvantages of such models are that (1) they lack biological realism (i.e., they require that fishing have an effect on the population within 1 year) and (2) they cannot make use of age- or size-specific information available from many fisheries. However, in some circumstances, surplus production models may provide better answers than age-structured models (Ludwig and Walters, 1985, 1989).

Delay-Difference or Aggregate-Matrix Models

These models incorporate age structure and provide a method for fitting an age- or size-structured population model to data aggregated by age (Deriso, 1980; Schnute, 1985; Horbowy, 1992). Estimation can be accomplished by fitting nonlinear model predictions of aggregate quantities to CPUE, biomass indices, and/or recruitment indices. Delay-difference models are a special-case solution to a more general aggregate-matrix model made possible by the assumption of a particular age-specific growth model (the von Bertalanffy equation [Ricker, 1975]). These types of models share the advantages of surplus production models; additionally, the functional relationship between productivity and abundance accounts for both yield-per-recruit and recruit-spawner effects. Unlike production models, the parameters of delay-difference or aggregate-matrix models have direct biological interpretations, but they cannot make full use of age- or size-specific information. In addition, these models require the estimation of more initial conditions than production models, unless a simplifying assumption, such as an initial equilibrium condition, is made.

Age-Based or Integrated Models

Age-based models use recursion equations to determine abundance of year classes as a function of several parameters (Fournier and Archibald, 1982; Deriso et al., 1985; Megrey, 1989; Methot, 1989, 1990; Gavaris, 1993). Relationships between spawning stock biomass and recruitment are not required but can be used. Because of the flexible nature of such general models, integration of many aspects of the data collection process with the population model is feasible. Estimation can be accomplished by maximum likelihood or least-squares procedures applied to age-specific indices of abundance, age-specific catch, and other types of auxiliary information. The main advantage of these models is that they make almost full use of available age-specific information. The primary disadvantage is that such models require many observations and include many parameters, increasing the cost of using them. ADAPT, Stock Synthesis, and CAGEAN/Cansar* belong in this category. *ADAPT* is an age-structured assessment method based on least-squares comparison of observed catch rates (generally age specific) and those predicted by a tunable† virtual population analysis. *Stock Synthesis* is an age-structured assessment technique based on maximum likelihood methods, but with more flexibility to include auxiliary information and fitting criteria. Additional details about ADAPT and Stock Synthesis are provided in Chapter 5.

Length- and Age-Based or Fully Integrated Models

Models that integrate length and age data are more complicated than age-based models because growth must be specified to relate length to age (Fournier and Doonan, 1987; Schnute, 1987; Deriso and Parma, 1988; Methot, 1990; Sullivan, 1992). This specification is accomplished with a stochastic-growth or length-age matrix conversion. Age structure of these models can be either implicitly or explicitly represented. Because of the flexible nature of such general models, full integration of the data collection process with the population model is feasible. Estimation can be accomplished by maximum likelihood procedures applied to age- and size-specific indices of abundance, age- and size-specific catch, and other types of auxiliary information. The main advantage of this type of model is that it makes full use of both age- and size-specific information. The primary disadvantage is that such models may require many observations and include many parameters.

In terms of data demands and number of parameters, these methods rank as follows (from least to most intensive): production models → delay-difference models → size-based models → age-based (or integrated) models → length- and age-based (or fully integrated) models. As models become more data intense and complex, there is a decreasing chance of gross model misidentification, an increasing chance of misidentifying some model component, increasing biological realism, and increasing cost of data collection.

Any stock assessment model involves choices at two levels: (1) the structural model that will be used and (2) the parameter values to be used. The choice of models is usually based on past experience. For example, models that assume the stock is in equilibrium are now almost universally avoided because experience has indicated that this assumption is rarely true; equilibrium models tend to produce results that are biased toward optimistic assessments of a stock's productivity. Any model used in an assessment includes many parameters that are assigned based on data other than those used in the assessment. The following are some examples of such parameters:

1. Natural mortality rate is usually assigned a fixed value based on a relationship to maximum age of the species, growth rate, maximum size, or other demographic or life history information (Vetter, 1988).

2. The relationship between indices of abundance and true abundance is generally assumed to be linear (as explained in Chapters 2 and this chapter), and the slope is usually estimated as part of the stock assessment.

3. Recruitment variability is most commonly left unconstrained, but it may be constrained to some value estimated from other data or based on experience.

4. A specific functional form for the relationship between stock and recruitment may be chosen and its

---

*CAGEAN (Deriso et al., 1985) is an age-structured assessment method based on forward-recursion population equations, a least-squares objective function (although other objective functions are also described), and lognormal distributions for catch-age and fishing mortality. An associated computer program provides estimates of population parameters and standard errors through a bootstrap routine. The algorithm is straightforward and can be implemented in spreadsheets and other computer software.

†Tuning of a model involves adjusting parameter estimates to minimize differences between predicted population estimates and observations from indices of population (e.g., catch rate, survey index of abundance).

parameters estimated jointly with abundance. Other sources of information about stock recruitment parameters can be included in the analysis, although this is not generally done.

5. Depensation[*] usually is assumed not to occur. Many assessment methods require no specification of this parameter.

USING SURVEY DATA IN MODELS

If properly calibrated, fishery-independent trawl surveys can be used to estimate the absolute abundance of a fish population. Numbers at age a in year t $(N_{a,t})$ can be estimated as

$$N_{a,t} = \rho_a \frac{D}{d} I_{a,t} \tag{3.2}$$

where ρ_a is the probability that a fish of age a in the path of the trawl is captured, D is the area of the survey stratum, d is the area swept by the trawl gear, and $I_{a,t}$ is the survey index of numbers at age.

For fixed-gear recovery methods (e.g., longlines, pots, gillnets) it is not possible to estimate the area sampled, and even for trawl surveys it is difficult to estimate the probability of capture (ρ_a) accurately. In these cases, estimates from the survey data are assumed to measure relative abundance and are combined with other information about the fish population to estimate existing and past population sizes. In the ADAPT model's original form, the survey estimate of population numbers at age is assumed to be related to actual population numbers as

$$I_{a,t} = q_a N_{a,t}^{\beta} e^{\varepsilon_{a,t}} \tag{3.3}$$

The parameter β is not defined in the model document, but is a commonly used form to express a nonlinear relationship between true abundance and a survey index; q_a represents the catchability of fish of age a to the survey gear and is assumed to be constant over time. The random component of the model $(\varepsilon_{a,\,t})$ is assumed to be an independent, symmetrically distributed random variable with constant variance and a mean value of zero. This distributional assumption allows for the use of standard nonlinear least squares to estimate the model parameters (Gavaris, 1993). Recently, Myers and Cadigan (1995) developed a random-effects mixed model within a maximum likelihood framework to estimate the parameters of the ADAPT model. Their approach allows for a correlated within-year error structure for the survey data to deal with "year effects" in survey abundance-at-age estimates (see Smith and Page, 1996). The implicit parameters of Equation (3.3) that must be estimated, whatever the approach, are the fishing mortalities used to estimate N_a in the underlying VPA (virtual population, or cohort, analysis) of the ADAPT method from catch-at-age data (Mohn and Cook, 1993). Absolute abundance is sometimes derived by calculating the area swept by the gear and assuming that all animals in the path of the gear will be captured (i.e., that catchability equals 1). In reality, absolute abundance estimates may be under- or overestimates, depending on whether catchability is greater than 1 due to herding by the trawl gear or less than 1 because of escapement from the path of the trawl.

A different formulation is used in the Stock Synthesis method (Methot, 1990), which more naturally accommodates year effects in the surveys. Instead of fitting the model to age-specific abundance indices, the Stock Synthesis model treats the indices of overall stock abundance separately from the age composition of the survey catches. Thus, year effects affect only year-specific abundance indices and do not introduce correlations among the age-specific observations. The expected value for survey numbers in year t (I_t) for the Stock Synthesis model (Methot, 1990) is determined as

---

[*]Depensation is a reduction in per capita productivity at low stock sizes.

$$I_t = q \sum_a s_a N_{a,t} \qquad (3.4)$$

where q is catchability for fully-recruited ages, s_a is age specific availability or selectivity, and $N_{a,t}$ is population abundance in numbers of fish in year t and age a. The survey can measure either relative ($q \neq 1$) or absolute ($q = 1$) abundance. Note the correspondence with the ADAPT formulation in (3.3) by letting $q_a = qs_a$ and $\beta = 1$.

In the Stock Synthesis method, variances from the log transform of survey abundance estimates are included, if available, directly in the survey index component of the log-likelihood expression. Therefore, the impact of optimizing the survey design on the resultant estimates from Stock Synthesis can be studied directly. Although there is the possibility of using inverse variance weighting in the nonlinear least squares in ADAPT, this is not usually done. However, bootstrap and Monte Carlo methods are available for linking variation in the survey estimates to variation in the resultant estimates from ADAPT (Restrepo et al., 1992; Smith and Gavaris, 1993b). Ultimately, sampling programs should be evaluated with respect to the precision of the quantities being used to estimate stock status.

The assumption of constant catchability and availability with age over time implied by the use of q_a in ADAPT and s_a in Stock Synthesis can be confounded by changes in both availability and catchability.

BAYESIAN APPROACHES

Fishery management involves decisionmaking in the presence of uncertainty. Fishery stock assessments should provide the quantitative support needed for managers to make regulatory decisions in the context of uncertainty. This support includes an evaluation of the consequences of alternative management actions. However, there is often considerable uncertainty that can be expressed as competing hypotheses about the dynamics and state of a fishery. The consequences of management actions may differ depending on which of these hypotheses is true. An appropriate means of providing quantitative support to managers in the presence of uncertainty is through the use of Bayesian statistical analysis. This section discusses the problem of building models of complex fishery systems having many parameters that are unknown or only partially known and the use of Bayesian methodology. Numerous papers and books have been published related to the application of Bayesian analyses in fisheries (e.g., Gelman et al., 1995; Punt and Hilborn, 1997).

There are three major elements in the Bayesian approach to statistics that should be indicated clearly:

1. likelihood of describing the observed data,
2. quantification of prior beliefs about a parameter in the form of a probability distribution and incorporation of these beliefs into the analysis, and
3. inferences about parameters and other unobserved quantities of interest are based exclusively on the probability of those quantities given the observed data and the prior probability distributions.

In a fully Bayesian model, unknown parameters for a system are replaced by known distributions for those parameters observed previously, usually called *priors*. If there is more than one parameter, each individual distribution, as well as the joint probability distributions, must be described.

A distinction must be made between Bayesian models, which assign distributions to the parameters, and Bayesian methods, which provide point estimates and intervals based on the Bayesian model. The properties of the methods can be assessed from the perspective of the Bayesian model or from the frequentist [*] perspective. Historically, the "true" Bayesian analyst relied heavily on the use of priors. However, the modern Bayesian has evolved a much more pragmatic view. If parameters can be assigned reasonable priors based on scientific knowledge, these are used (Kass and Wasserman, 1996). Otherwise, "noninformative" or "reference" priors are

---

[*]Frequentist statistical theory measures the quality of an estimator based on repeated sampling with a fixed, nonrandom set of parameters. Bayesian statistical theory measures the quality of an estimator based on repeated sampling in which the parameters also vary according to the prior distributions. Most beginning statistics courses focus on frequentist methods such as the *t*-test and analysis of variance.

used.* These priors are, in effect, designed to give resulting methods properties that are nearly identical to those of the standard frequentist methods. Thus, the Bayesian model and methodology can simply be routes that lead to good statistical procedures, generally ones with nearly optimal frequentist properties. In fact, Bayesian methods can work well from a frequentist perspective, as long as the priors are reasonably vague about the true state of nature. In addition to providing point estimates with frequentist optimality properties, the posterior intervals for those parameter estimates are, in large data sets, very close to confidence intervals. Part of the modern Bayesian tool kit involves assessing the sensitivity of the conclusions to the priors chosen, to ensure that the exact form of the priors did not have a significant effect in the analysis. There are differences of opinion among scientists about whether frequentist or Bayesian statistics should be used for making inferences from fishery and ecological data (Dennis, 1996).

The basis for selection of the various prior distributions used in a stock assessment should be documented because the choice of priors can be a source of improper use of statistics. A rationale has to be constructed to indicate which models were considered in the analysis and why some models were not considered further (i.e., were given a prior probability of zero), even though they may be plausible. The use of standard procedures permits independent scientific review bodies to verify the plausibility of hypotheses included in the assessment and assign their own prior probability values to the selected hypotheses.

There are two general classes of Bayesian methods. Both are based on the posterior density, which describes the conditional probabilities of the parameters given the observed data. This is, in effect, a modified version of the models' prior distribution, where the modification updates the prior based on new information provided by the data. In one form of methodology, this posterior distribution is maximized over all parameters to obtain "maximum a posterori" (MAP) estimators. It has the same potential problem as maximum likelihood in that it may require maximization of a high-dimension function that has multiple local maxima. The second class of methods generates point estimators for the parameters by finding their expectations under the posterior density. In this class, the problem of high-dimension maximization is replaced with the problem of high-dimension integration.

Over the past 10 years, Bayesian approaches have incorporated improved computational methods. Formerly, the process of averaging over the posterior distribution was carried out by traditional methods of numerical integration, which became dramatically more difficult as the number of different parameters in the model increased. In the modern approach, the necessary mean values are calculated by simulation using a variety of computational devices related more to statistics than to traditional numerical methods. Although this can greatly increase the efficiency of multiparameter calculations, the model priors must be specified with structures that make the simulation approach feasible.

Although they are not dealt with extensively here, a number of classes of models and methods have an intermediate character. For example, there are "empirical Bayes" methods, in which some of the parameters are viewed as arising from a distribution that is not completely known but rather known up to several parameters. There are also "penalized likelihood methods," in which the likelihood is maximized after addition of a term that avoids undesirable solutions by assigning large penalty values to unfeasible parameter values. The net effect is much like having a prior that assigns greater weight to more reasonable solutions, then maximizes the resulting posterior. Another methodology used to handle many nuisance parameters is the "integrated" likelihood, in which priors are assigned to some of the parameters to integrate them out while the others are treated as unknown. This provides a natural hybrid modeling method that could have fishery applications.

Advantages of Bayesian Models and Methods

A number of features of Bayesian modeling make it particularly useful for fish stock assessments:

1. In a complex model, if a key parameter is treated as totally unknown, the set of parameters of the model

---

*Such priors are sometimes "improper" in that the specified prior density is not a true density because it does not integrate to 1. A prior distribution is proper if it integrates to 1.

may become nonidentifiable (i.e., parameters cannot be estimated consistently no matter how many data are collected). On the other hand, treating a parameter as completely known is too optimistic. Assigning the parameter a known distribution can provide a compromise of including the uncertainty about its true value while retaining the identifiability of the remaining parameters. For example, the natural mortality rate is required in most fishery models but can seldom be estimated accurately. Instead of assuming a known value, natural mortality could be included as a distribution.

2. Fish stock assessment takes place in an environment of decisionmaking. It involves evaluating possible consequences of alternative actions across competing hypotheses about the system state and its dynamics. Bayesian models are appropriate because they provide probabilities that each of these hypotheses is true, conditional on available information. A distinction can be made between statistical methods designed to provide an estimate based on information in the data alone and those that are designed to use all available information optimally. Bayesian methods allow one to do the latter, following the spirit of National Standard #2 of the Magnuson-Stevens Fishery Conservation and Management Act, which calls for "conservation and management measures . . . based upon the best scientific information available" (16 U.S.C. 1851).

3. Meta analysis, increasingly used in fisheries science, provides information in an ideal form for Bayesian analysis. For example, if a particular fishery parameter has a distribution over a class of similar species that have been studied more completely, with appropriate adjustment a ready-made prior distribution is available for that parameter in the species of interest.

4. The use of Bayesian models addresses, to some extent, under- versus overparameterization. A model with too few parameters is most likely to provide biased estimators, but with the least variability in its individual estimators. A model with many parameters can fit the actual population dynamics very well but may vary so much over repeated runs that it provides poor estimation. If the larger model has its parameters modeled with informative priors, it lies somewhere between these two options and can be a desirable compromise. A simple model often produces better point estimates merely because the estimates are more stable and less variable, and have some bias that may be insignificant compared to variation. However, a simple model is doomed to underestimate variability. The advantage of a complex model is that all uncertainty is present in the width of the confidence statements rather than hidden.

5. With a completely specified probability model that allows for future uncertainties, simulation studies can provide management strategies that cope with the variability in future fisheries. Forecasting can allow for increasing uncertainty about the future values of input parameters as well as increasing knowledge of the system's fixed parameters.

6. Many complex fishery models are difficult to fit because the likelihood or least squares problems are highly nonlinear. Thus, modeling approaches require the use of multiple starting values and tuning parameters. However, some Bayesian methods are based not on finding maxima or minima but rather on finding posterior means, generally by a process of averaging over separate computer runs. Clearly, this method is more stable as well as more amenable to independent verification, and it is quite repeatable when similar priors are used.

Limitations of the Bayesian Method

The case for Bayesian methods presented above must be tempered with some limitations of the methods. Several important issues of which the user should be aware include the following:

1. Specification of the priors in a Bayesian model is an emerging art. To do a fully Bayesian analysis in a complex setting, with no reasonable prior distributions available from scientific information, requires a careful construction whose effect on the final analysis is not clear without sensitivity analysis. If some priors are available but not for all parameters, a hybrid methodology, which does not yet fully exist, is preferred.

2. In a complex model, it is known by direct calculation that the Bayesian posterior mean may not be consistent in repeated sampling. However, little information is available to identify circumstances that could lead to bad estimators.

3. Although modern computer methods have revolutionized the use of Bayesian methods, they have created

some new problems. One important practical issue for Markov Chain-Monte Carlo methods is construction of the stopping rule for the simulation. This problem has yet to receive a fully satisfactory solution for multimodal distributions. Additionally, if one constructs a Bayesian model with noninformative priors, it is possible that even though there is no Bayesian solution in the sense that a posterior density distribution does not exist, the computer will still generate what appears to be a valid posterior distribution (Hobert and Casella, 1996).

4. There is a relative paucity of techniques and methods both for diagnostics of model fit, which is usually done with residual diagnostics or goodness-of-fit tests, and for making methods more robust to deviations from the model. Formally speaking, a Bayesian model is a closed system of undeniable truth, lacking an exterior viewpoint to make a rational model assessment or to construct estimators that are robust to the model building process. To do so, and retain the Bayesian structure, requires constructing a yet more complex Bayesian model that includes all reasonable alternatives to the model in question, and then assessing the posterior probability of the original model within this setting.

5. Despite the progress in numerical integration of posterior densities, models with large numbers of parameters can be difficult to integrate. Two of the most commonly used methods, (a) sampling importance resampling and (b) Monte Carlo-Markov Chain, each have difficulty with multimodal or highly nonelliptical density surfaces.

META ANALYSIS

Meta analysis is broadly defined as a quantitative method for combining information across related, but independent, studies. The motivation for meta analysis is to integrate information over several studies and to summarize the information. Hedges and Olkin (1985) provide a good explanation of the statistical methods involved. This technique has been applied extensively to biomedical data (D'Agostino and Weintraub, 1995; Marshall et al., 1996) where results from different laboratories and experiments are combined. As applied to stock assessments, meta analysis involves the compilation of preexisting data sets to determine the values of parameters of models or to develop prior probability distributions for these parameters. The most widely used meta analysis method for fisheries was described by Pauly (1980), who examined the relationship among natural mortality rate, water temperature, and growth parameters for a large number of fish stocks. Myers et al. (1994) performed a meta analysis for sensitivity of recruitment to spawning stock, and Myers et al. (1995) conducted a meta analysis of depensation. There are two recent developments in meta analysis for stock assessments: (1) the compilation of large data sets, for example, by Myers and his colleagues; and (2) the application of formal Bayesian methods to estimate prior probability distributions. This allows the use of what is known from other stocks and species to put limits or distributions on parameters and thus obtain a realistic estimate of uncertainty for a new stock or species.

Many parameters are used in stock assessment models, including natural mortality, catchability, potential nonlinearity between indices of abundance, and actual abundance. The general practice has been to assume that some parameters are known perfectly without error. If, however, we were honestly to assess our uncertainty about these parameters, the uncertainty in the overall stock assessment would be large, in some instances, so large as to make results meaningless.

The underlying assumption behind meta analysis is that a parameter is replaced with a random variable. For instance, an estimation of the natural mortality rate among stocks of cod should exhibit a distribution of values. In the absence of any other data for the stock, the distribution from all cod stocks could be used as the probability distribution for the particular cod stock of interest. The distribution among all populations is called the *hyperdistribution*. The simplest approach to estimating hyperdistribution is to plot the frequency distribution of parameter estimates available for all stocks of interest. The problem with such empirical frequency distributions is that each estimate of the parameter for a particular stock involves sampling error. Thus, the empirical distribution is expected to have a higher variance than the true hyperdistribution. A method known as hierarchic Bayesian analysis has been developed to deal with this problem and is described in Gelman et al. (1995). This method has been used by Liermann and Hilborn (in press) to develop probability distributions for depensation in spawner-recruit relationships. Eddy et al. (1992) describe a new set of meta analysis techniques known as the confidence profile method, which may be applicable to some fisheries problems.

There are a number of potential problems in meta analysis. Particularly important is the fact that species or stocks represented in the meta analysis could show patterns of selection bias that would make them unrepresentative (in the variable of interest) within the spectrum of possible species. For example, species of high economic value or large population size are more likely to have been studied in the past, but they could differ in key biological characteristics from the new species of interest.

RETROSPECTIVE ANALYSIS IN STOCK ASSESSMENTS

The reliance of most stock assessment models on time-series data implies not only that each successive assessment characterizes current stock status and other parameters used for management, but also that the complete time series of past abundance estimates is updated. Retrospective analysis is the examination of the consistency among successive estimates of the same parameters obtained as new data are gathered. Either the actual results from historical assessments are used or, to isolate the effects of changes in methodology, the same method is applied repeatedly to segments of the data series to reproduce what would have been obtained annually if the newest method had been used for past assessments.

Retrospective analysis has been applied most commonly to age-structured assessments (Sinclair et al., 1991; Mohn, 1993; Parma, 1993; Anon., 1995b). In such applications, the statistical variance of the abundance (or fishing mortality) estimates tends to decrease with time elapsed; estimates for the last year (those used for setting regulations) are the least reliable. In retrospective analysis, abundance estimates for the final years of each data series can vary substantially among successive updates, whereas those for the early years tend to converge to stable values. In some cases (e.g., some Northwest Atlantic cod stocks, Pacific halibut, North Sea sole), early abundance estimates are consistently biased (either upward or downward) with respect to corresponding estimates obtained in later assessments. Extreme cases of consistent overestimation of stock abundance can have disastrous management consequences, as illustrated by the collapse of the Newfoundland northern cod (Hutchings and Myers, 1994; Walters and Pearse, 1996).

Retrospective biases can arise for many reasons, ranging from bias in the data (e.g., catch misreporting) to different types of model misspecification (mostly parameters that are assumed to be constant in the analysis but actually change, as well as incorrect assumptions about relative vulnerability of age classes). In traditional retrospective analyses, successive assessments use data for different periods, all starting at the same time with one year of data added to each assessment. An alternative method is to conduct successive assessments using data for a moving window of a fixed number of years (as in Parma, 1993, and Deriso et al., 1985). This method is appropriate for exploring trends in parameter estimates.

Ad hoc adjustment factors based on consistent past retrospective errors have been applied occasionally to correct the estimates used for management (e.g., Showell and Bourbonnais, 1994). For example, if past estimates of abundance were shown to be about 40% above the revised estimates obtained subsequently, current estimates could be adjusted downward to compensate, in the expectation that similar bias would also be present in the last year's estimate. Although ad hoc, this may be a sensible precaution in cases of historical overestimation of abundance. Conversely, adjusting current abundance estimates upward to compensate for negative retrospective bias is risky, because the sign of the retrospective errors can reverse without warning; Pacific halibut illustrates such reversal (see Parma and Sullivan, 1996). Adams et al. (1997) have recently developed resampling tests to ensure proper evaluation of the main effects in meta-analysis applications.

Retrospective analysis is an effective tool for uncovering potential problems in an assessment methodology, even if it fails to provide clues about their possible sources. Some of these problems, such as trends in catchability affecting the reliability of the abundance data used for fitting the models, can also be detected as autocorrelation in the residual error terms of statistical analyses. However, the magnitude of the problem is difficult to assess from a single fit to a time series data set. Consistent patterns in retrospective errors indicate problems in the specification of the model; thus, conventional model-based measures of uncertainty of management parameters are not realistic because they are based on the structural assumptions of the model being correct. Because there are many possible sources of the retrospective problem, each specific case has to be considered individually in the search for solutions. An illustration of the use of retrospective analysis is shown in Chapter 5 in a presentation of the committee's analysis of the results of its simulations.

DATA WEIGHTING

A crucial problem that has emerged in modern stock assessments that combine multiple sources of data and information about population parameters is how to weight each type of information. The typical approach is to define a least-squares or likelihood objective function that contains weight parameters. If $\{X, Y, \ldots\}$ represent various data sources, X_i is an observation, $\hat{X}_i$ is the corresponding model estimate, and σ_x^2 is the variance of that observation, then the typical least-squares objective function is written

$$\text{SSQ} = \sum_i (X_i - \hat{X}_i)^2 / \sigma_x^2 + \sum (Y_i - \hat{Y}_i)^2 / \sigma_Y^2 + \ldots \tag{3.5}$$

where the data weighting is inversely proportional to variance (Deriso et al., 1985). Similar expressions can be written for likelihood or Bayesian approaches, and this method can be generalized for the situation in which variance changes from observation to observation.

In practice, these variances are unknown, and different approaches have been used to estimate them or to substitute weights $\{\Sigma_X, \Sigma_Y\}$ for the inverse of the variances. Deriso et al. (1985) found that for Pacific halibut data, weighting choice was unimportant for a midrange of weightings. However, many other studies and assessments since then have shown that parameter estimates are related monotonically to the weights chosen. Kimura (1990) showed an empirical approach for obtaining weights from sampling considerations. The variances in Equation (3.5) are often inversely proportional to sample size, suggesting that weighting by sample size is a reasonable approach. Many analysts have no a priori information about weighting, so they assign each observation equal weight. Others prefer to weight each data component equally when the values for different data components differ greatly. Some researchers suggest using perceptions about the data as weights, with either Bayesian or least-squares approaches (Geiger and Koenings, 1991; Merritt, 1995). This issue will be more broadly explored at the 1997 Lowell Wakefield Symposium sponsored by the Alaska Sea Grant College Program.

UNCERTAINTY IN STOCK ASSESSMENT METHODS AND MODELS

Stock assessments are intrinsically uncertain. Sources of uncertainty include (1) variability and nonstationarity of stock dynamics, (2) errors in data due to sampling variability, and (3) errors in model specification. The dynamics of fish stock growth, together with fluctuations in environmental conditions, result in stochastic variation in fish abundance. Many stock assessment methods and models in current use are homogeneous (deterministic) in the sense that parameters do not vary in relation to spatial or temporal variations in the environment.

A simple example of subjective uncertainty from fisheries might be an estimate of the instantaneous natural mortality rate of a stock of fish. It is not uncommon to have only one estimate of natural mortality rate derived from empirical data. There are many times when assessment analysts simply use their experience and judgment to assess a parameter value and ranges of feasibility for the parameter, rather than measuring the parameter directly. A Bayesian solution could hide the sensitivity of the analysis to assumptions about this parameter, whether estimated by a point value or a prior. Most committee members agreed that Bayesian approaches are the most promising means to build uncertainty into stock assessment models. However, at least one committee member is not convinced that more complex models (especially if they are extensions of present methods) are a rational solution and believes that novel approaches to deal with uncertainty should continue to be explored. Examples of such novel approaches include fuzzy arithmetic and interval analysis.

Interval analysis may be appropriate to propagate uncertain values through calculations. *Fuzzy numbers* represent a generalization of intervals in which the bounds vary according to the confidence one has in the estimations. Both interval analysis and fuzzy arithmetic may be useful approaches for propagating uncertainty in calculations under some conditions. Application of fuzzy arithmetic to fisheries is demonstrated by Ferson (1994) and Saila et al. (in press). Details about fuzzy arithmetic and interval analysis are given in Appendix K.

4

Harvest Strategies

In addition to estimates of stock size, fisheries stock assessments also provide decisionmakers with a quantitative evaluation of the consequences of alternative actions. Although the data and assessments discussed in Chapters 2 and 3 provide an estimate of the population size, recruitment rates, and other important values, the assessment process does not stop there. A biological representation of the stock dynamics must be incorporated in an evaluation of the consequences of alternative actions, encompassing the *harvest strategy*. The utility of assessment methods cannot be evaluated without considering how the assessments are used to choose among alternative harvest strategies.

Three related terms must be distinguished:

1. *Harvest strategies* are the plans for adjusting management options in relation to the status of the fish stock. The two most common harvest strategies are (a) fixed exploitation rate, in which an attempt is made to take a constant fraction of the fish stock each year, and (b) constant escapement, in which an attempt is made to maintain the spawning stock size near some constant level (Figure 4.1). Management of fish populations to sustain catches and abundance levels can be based on several alternative means of strategic catch regulation. Fixed exploitation rate strategies are commonly employed in marine fish stocks under such names as F_{MSY}, $F_{0.1}$, F_{max}, and $F_{40\%}$ (defined later in this chapter). Constant escapement strategies are less commonly used, primarily for many stocks of Pacific salmon.

2. *Harvest tactics* are the regulatory tools (e.g., quotas, seasons, gear restrictions) used to implement a harvest strategy. Harvest tactics are quite diverse, and almost all fisheries employ gear restrictions, area restrictions, and some limitation of seasons. Quotas are increasingly employed in large-scale commercial fisheries, whereas closed seasons and closed areas are common in recreational fisheries.

3. *Management procedures* represent the combination of data collection, assessment procedure, harvest strategy, and harvest tactics.

The majority of this chapter discusses how to evaluate alternative harvest strategies, although in principle it would be considerably better to evaluate the entire management procedure.

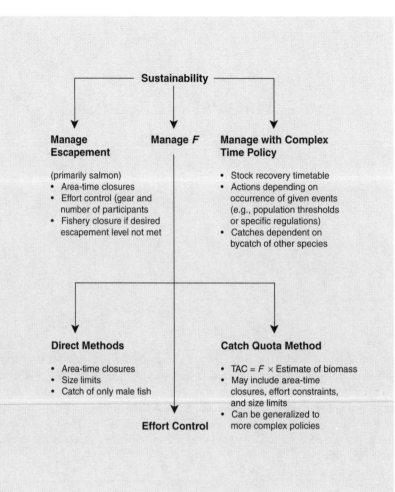

FIGURE 4.1   Role of stock assessment methods in sustainable fisheries.

INDICATORS OF PERFORMANCE

Evaluating alternative harvest strategies requires the definition of a suite of indicators to measure the expected performance of an entire fishery system including, for example, projections of fish stock size, the probability of dropping below certain thresholds, and projected catch per unit effort (CPUE). Empirical indices measure the historical performance of a fishery up to the present; analytical indices are used to project the future consequences of management decisions. In either case, the time frame of interest is typically short—about 10 years before and after the present, with greatest emphasis on present values. In *Our Living Oceans* (NMFS, 1996), for example, *recent average yield* includes only the most recent three-year period. The objectives of fishery management are seldom clearly articulated, but by noting which performance indicators are most often reported, the quantities with the greatest importance can be identified. The most common indicators involve estimates of the health of fish populations (biological indicators), the performance of the fishery (yield and social indicators), and indicators of how the uncertainty in key population parameters is likely to change in the future depending on current management actions (uncertainty indicators).

Biological Indicators

Biological indicators measure the status of a fish stock. The most widely reported indicators are trends in survey indices and trends in CPUE. For stocks that are managed using age-structured assessments, spawning stock biomass (SSB, the weight of reproductively mature fish) is an important indicator of reproductive capacity. The overfishing definitions for Atlantic mackerel, Gulf of Mexico shrimp, and northern anchovy are all based on maintaining minimum levels of SSB (Rosenberg et al., 1994). Likewise, the overfishing definitions for Pacific salmon are based on target numbers of spawning salmon. For some species, such as striped bass, a juvenile index is used to indicate the supply of recruits to the harvestable population (NMFS, 1996). Under satisfactory sampling conditions, juvenile indices can provide an effective measure of recruits (de Lafontaine et al., 1991), but problems with time-varying catchability hamper the usefulness of juvenile indices in other fishery assessments (e.g., Pacific halibut, see Quinn, 1985). The risks associated with low stock sizes include reduced fisheries landings, recruitment overfishing,[*] lack of forage fish for predators, and potentially irreversible changes to the ecosystem structure and function.

Yield and Social Indicators

Yield indicators measure the outputs of a fishery, namely, the recreational and commercial landings. Probably the most important yield indicator is the landed catch (*landings*) averaged over some time period. If the time frame of averaging is long enough to span several generations of the fish, landings also reflect the stock's SSB. Averaging is necessary to smooth out fluctuations in yield that result from environmentally driven variations in recruitment. In most stochastic models of fish populations, there is a trade-off between maximizing mean catch and minimizing the standard deviation of catch (e.g., Figure 9 in Collie and Spencer, 1993). A linear combination of yield (*Y*) and standard deviation of yield (SD) was used as an objective function by Quinn et al. (1990):

$$\max[(1-p)Y - pSD] \tag{4.1}$$

where *p* is a penalty weighting factor that represents the cost of one unit of increase in standard deviation in relation to one unit of increase in average yield.

Projected landings are often *discounted* to reflect the fact that present landings have a greater economic value than fish caught some time in the future. The present value of landings (*PV*) caught *t* years in the future is compared to the landings caught at the present time (*C*):

$$PV = \frac{C}{(1+i)^t} \tag{4.2}$$

where $(1+i)^t$ is the discount factor (Clark, 1985). Although the appropriate value of the interest rate (*i*) is open to discussion, a relatively low discount rate of 2 to 5% is often used in fishery management studies (Walters, 1986). The choice of discount rate is important because it implicitly sets the planning horizon. The U.S. Office of Management and Budget (OMB) requires use of a 7% discount rate in the cost-benefit analyses that accompany fishery management plans.[†] If $i = 7\%$, then by $t = 68$ years, *PV* will drop to approximately 1% of present landings and landings beyond 68 years have almost no present value.

---

[*]Recruitment overfishing results from fishing at a high enough level to reduce the biomass of reproductively mature fish (*spawning biomass*) to a level at which future recruitment is impacted.

[†]OMB Circular A-94 at http://www.whitehouse.gov/WH/EOP/OMB/html/circulars, June 25, 1997.

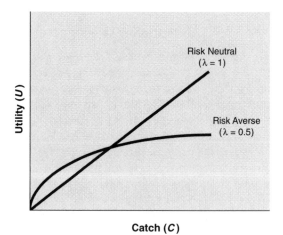

FIGURE 4.2   Fisheries utility functions.

Variability in annual landings is generally considered undesirable compared to maintaining the average yield at a constant level because variable landings create uncertainty in fishing communities and lead to the inefficient and intermittent use of capital by fishers. In fisheries with threshold harvest policies, an important yield indicator is the number of years in which the stock falls below the threshold, closing the fishery (Hall et al., 1988; Quinn et al., 1990). This penalty against allowing the stock to drop too low can be expressed more generally with the *risk-averse* utility function, $U = C^\lambda$ (Figure 4.2), where $0 < \lambda < 1$ (Mendelsson, 1982). With this type of concave utility function, the penalty associated with landings dropping below their mean value outweighs the benefits associated with a similar increase in landing above the mean value. A special case of the risk-averse utility function, $U = \log(C)$, has been used in studies of harvest policy (Walters and Ludwig, 1987), but it has the disadvantage of being undefined if the fishery is closed ($C = 0$).

The risk-averse utility function was originally formulated to describe the situation in which price is assumed to decrease as the quantity of landings increases (Clark, 1985). In this case, the benefits of increased landings above the mean level are offset by the associated decrease in price. Data on fish prices are widely available, and demand models exist that can be used to project prices as a function of harvest. The landed value (landed weight of fish in kilograms times price per kilogram) is a common performance indicator that represents the gross benefits from a fishery. Ideally, the net benefits can be calculated by subtracting the costs of catching the fish, managing the fishery, and other related activities. In practice, fishing costs are difficult to measure and net benefits can seldom be estimated reliably. In some fisheries with a risk-averse utility function, when price per unit declines as landings increase, any decrease in TAC to help increase price per unit is lost if the landings are made up through imports. This actually exacerbates, rather than relieves, the economic problems for fishers. Some yield indicators can also function as social indicators, such as net economic benefit from a fishery. Other social indicators are employment in the fishery and associated industries and average debt burden of individual fishers and fishing companies.

Uncertainty Indicators

A final class of performance indicators measures the rate at which analysts can learn about uncertain population parameters. Some harvest policies may be more informative than others in reducing parameter uncertainty or allowing the analyst to choose among a set of competing hypotheses about stock production (Walters, 1986). It follows that a more informative harvest policy should lead to better management and increased future benefits. Four measures of learning performance were discussed in detail by Walters (1986). In brief, high performance is associated with minimizing the uncertainty about key population parameters and policy variables and minimizing the error in predicting future stock sizes.

BIOLOGICAL REFERENCE POINTS

The 1992 guidelines for fishery management plans (50 CFR, Part 602) stipulate that overfishing definitions must exist for all stocks managed under federal fishery management plans. For this reason, much effort has been devoted to defining overfishing thresholds (Rosenberg et al., 1994). Biological reference points (BRPs) are calculable quantities that describe a population's state. They can be used as targets for optimal fishing, as well as for setting overfishing thresholds. They are calculated from the life-history characteristics of a given stock and are used to define harvest control laws (Figure 4.3). A BRP can be expressed as a fishing mortality rate (F) and/or as a level of stock biomass (B). BRPs can be targets or thresholds.

A threshold control rule specifies the upper limit of fishing mortality allowable or the lower biomass limit beyond which overfishing occurs. A target control rule is more conservative than a threshold and defines a desired rate of fishing and acceptable levels of stock biomass. It is wise to have some separation between the target and threshold levels, so that minor overruns of targets will not exceed the thresholds. For many depleted stocks, the overfishing thresholds have become de facto targets, contrary to the intent of the overfishing definitions, leaving no buffer to accommodate occasional overestimates or unexpected negative environmental factors. For example, pollution events, disease or predator outbreaks, and unusually warm or cold water temperatures can affect fish stocks unexpectedly. More complicated strategies involving yield per recruit (e.g., eumetric fishing [Beverton and Holt, 1993]) have been considered but have been superseded for the most part by direct consideration or more detailed modeling of selectivity and catchability. Annex II of the Agreement for the Implementation of the United Nations Convention on the Law of the Sea of 10 December 1982 Relating to the Conservation and Management of Straddling Fish Stocks and Highly Migratory Fish Stocks (which the United States has ratified) provides guidelines for applying precautionary reference points for managing populations of highly migratory fish species and those that straddle national boundaries (UN, 1995).

Fishing Mortality Reference Points

The natural mortality rate (M), or some fraction of M, has been used in some fisheries to set the F value that would constitute overfishing (e.g., for North Pacific groundfish). The rationale for this approach is that short-lived species with high M (and presumably a high intrinsic rate of increase) should be able to sustain a higher F level

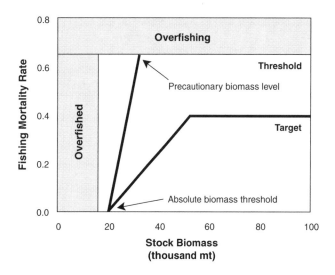

FIGURE 4.3 Schematic harvest control laws.
SOURCE: Modified from Rosenberg et al. (1994).
NOTE: mt = metric tons.

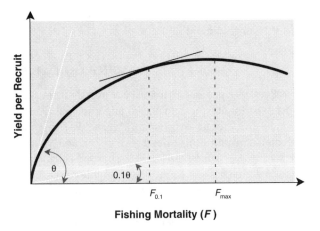

FIGURE 4.4   Method of defining $F_{0.1}$, given a known relationship between fishing mortality rate and yield per recruit, as the point on the *Y/R* curve at which the slope of a line tangential to the curve is one-tenth the slope of a line tangential to the curve at the origin.

than long-lived species with low *M*. Indeed, many fish stocks that have sustained fisheries for long periods have sustained fishing mortality rates near *M* (Mace, 1994). Unfortunately, *M* is not well known for most fish stocks, and estimated values must be used in stock assessments. Two common *M*-based reference points use the yield per recruit (YPR) or spawning biomass per recruit (SPR).

Yield per Recruit

YPR is the total yield in weight harvested from a year-class of fish over its lifetime, divided by the number of fish recruited into the stock (Figure 4.4). The relationship between YPR and *F* is useful for defining several BRPs. The threshold level F_{max} is the fishing mortality that maximizes YPR. Levels of fishing mortality higher than this reference point constitute *growth overfishing* because individuals are harvested before they have grown to a size that will maximize YPR. F_{max} is used as an overfishing definition for summer flounder and scup in the northeastern United States. It is possible to calculate the age of first entry to the fishery (or mesh size) that will result in a given YPR with a minimum expenditure of fishing effort.

Another reference level ($F_{0.1}$) is defined as the *F* at which the slope of a line tangential to the YPR curve is one-tenth of its slope at the origin (Gulland and Boerma, 1973; Figure 4.4). Despite its arbitrary definition, $F_{0.1}$ is desirable because it is a target reference point that is lower than F_{max} and therefore provides a buffer to avoid growth overfishing. The $F_{0.1}$ level increases yield per unit effort without sacrificing much yield and has been widely used as a target *F* (Deriso, 1987). $F_{0.1}$ also has some basis in bioeconomics because it is the point at which each additional unit of fishing mortality achieves less than 10% of the yield per recruit obtainable from a unit of fishing mortality applied to a previously unexploited stock; that is, the return (in units of catch biomass per recruit) on investment in a unit of fishing mortality is 10% of the return obtainable from the stock when it was in an unexploited condition. The product of equilibrium recruitment (*R*) and YPR can be used to define F_{MSY}, the fishing mortality for maximum sustainable yield (MSY; Figure 4.5).

Spawning Biomass per Recruit

A common SPR-based reference level, $F_{X\%}$, is defined as the *F* that would reduce the spawning stock biomass per recruit to X% of the level that would exist with no fishing. Many of the overfishing definitions for U.S. fish stocks are based on an $F_{X\%}$ (Rosenberg et al., 1994) because it relates directly to SSB, the quantity to be conserved. Specifying the $F_{X\%}$ to use as a reference level is somewhat arbitrary; recent work has sought to define $F_{X\%}$ values that are analogous to $F_{0.1}$. For example, Clark (1993) and Mace and Sissenwine (1993) suggest that levels of $F_{35\%}$ to $F_{45\%}$ are appropriate, as explained below. Spawning biomass per recruit is complicated in

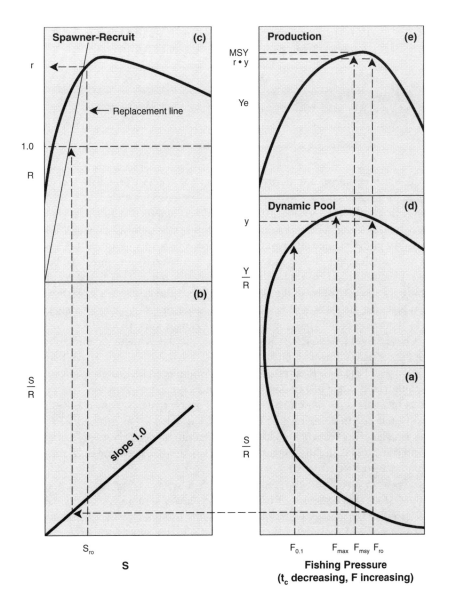

FIGURE 4.5   Single species theory of fishing. A dynamic pool model (a and d) describes the effect of fishing mortality rate (F) and age at capture (t_c) on spawning biomass (S) per recruit (R) and yield (Y) per recruit. A spawner-recruit model (c) relates the number of recruits to spawning biomass; a "replacement line" with slope equal to the inverse of S/R is mapped by a graphic procedure (b). S and R must be scaled to appropriate units (e.g., thousands of tons and millions of fish, respectively) for the graphic procedure to be practical. The intersection of the replacement line with the spawner-recruit function determines equilibrium recruitment. The product of equilibrium recruitment (c) and yield per recruit (d) is the equilibrium production (e). The method is demonstrated for a fishing mortality rate of F_{ro} (conditional on an unspecified value of t_c). F_{MSY} is indicated based on the production model (c); $F_{0.1}$ and F_{max} are indicated in (d). SOURCE: Sissenwine and Shepherd (1987). Used with permission from Canada's National Research Council Press.

TABLE 4.1   Relationships Between Fishing Mortality
Reference Levels and Spawning Biomass Per Recruit

BRP	Corresponding Average %SPR	Reference type
$F_{0.1}$	38	Target
F_{max}	21	Threshold
F_{med}	19	Threshold

hermaphroditic fish, which comprise a minority of commercial species, primarily reef dwellers in tropical and semitropical regions.

Shepherd (1982) first showed how SPR analysis could be combined with stock-recruitment (S-R) observations[*] to generate reference fishing mortality rates. An example is F_{med}, the fishing mortality rate that results in an SPR equal to the inverse of the median survival ratio (number of recruits produced per unit of spawning biomass [R/SSB]). Sissenwine and Shepherd (1987) suggested F_{med} as an overfishing threshold because a stock harvested at this rate (or lower) should be able to replace itself on average, but F_{med} is based on observed survival ratios, which depend on the exploitation history of the stock. If F_{med} is estimated after a stock has been exploited heavily, its value will be too high and the stock will be depleted. Conversely, F_{med} may be a very conservative threshold in the case of stocks that are exploited only lightly before it is determined.

Several studies have examined the BRPs among stocks, and the mathematical relationships among BRPs, to determine the most appropriate target and threshold levels (Deriso, 1987). For 91 well-studied stocks, Mace and Sissenwine (1993) computed the %SPR corresponding to fishing mortality reference levels (Table 4.1).

Clark (1991) examined a range of life-history parameters typical of demersal fish to determine the level of fishing mortality that maximizes the minimum yield (F_{mmy}) over a range of typical S-R relationships (F_{mmy} is the best fishing mortality rate for minimizing lost harvest). Clark found that in general, $F_{mmy} \approx F_{35\%} \approx F_{0.1}$ and advocated $F_{35\%}$ as a target F, except for stocks where recruitment and maturity schedules do not coincide. Subsequent trials indicated that a slightly higher %SPR of about 40% is desirable in the presence of high serial correlation in recruitment from year to year (Clark, 1993).

Many species appear to have strongly compensatory S-R relationships; that is, per capita recruitment increases significantly as stock size decreases. Reference levels are now more commonly based on a %SPR, but the percentage is often specified by analogy with other stocks or by using the results mentioned above. A knowledge of the compensatory capacity of the stock is necessary to define the most appropriate BRPs for a stock. Even without such knowledge, however, a conservative %SPR still can be selected (Sissenwine and Shepherd, 1987).

Biomass-Based Reference Points

Biomass-based reference points may be defined as sharp thresholds or as sliding scales to reduce fishing mortality as stock size declines progressively. Observed stock-recruitment data can be used to define biomass thresholds. Biomass thresholds can also be defined from fitted S-R relationships. An example is the SSB corresponding to 50 percent of the maximum expected recruitment (Mace, 1994). These estimators are easily understood and relatively robust if only data at low stock sizes are available and almost always result in higher levels of recruitment above the minimum biomass threshold (Myers et al., 1994).

A biomass threshold can be defined as a percentage of the estimated unexploited stock size (B_0), for example, $20\%B_0$ (Beddington and Cooke, 1983; Quinn et al., 1990). The appropriate percentile depends on the compensatory capacity of the stock; stocks with higher compensatory capacity can sustain a lower biomass threshold. B_0 must be estimated either with a stock-production relationship or by assuming that the earliest recorded biomass represented an unfished condition. Finally, stock-production relationships can be used to estimate B_{MSY}, which

---

[*]Stock-recruitment relationships quantify the amount of recruitment that occurs for a species, given a specific spawning biomass.

may be considered a target biomass level, the biomass at which maximum sustainable yield could be attained. As mentioned earlier, Atlantic mackerel, Gulf of Mexico shrimp, and northern anchovy stocks are managed using biomass-based BRPs. The North Pacific Fishery Management Council recently established a default threshold of 5% of unfished biomass and a biomass-based adjustment to decrease F at low population sizes. Many Alaska salmon populations have target minimum biomass levels (in numbers of fish).

The most appropriate reference point chosen to manage harvest from a specific stock depends on how environmental variability affects the dynamics of the population. Environmental variability can affect population growth rates directly via density-independent rate processes (Beddington and May, 1977) or via density-dependent processes and equilibrium abundances (Roughgarden, 1975; Shepherd and Horwood, 1979). The latter case is consistent with variability at long time scales (Pimm, 1984) as is the case for marine fishes. Consider the Ricker model:

$$R = \alpha S e^{-\beta S} \tag{4.3}$$

If environmental variability affects the density-independent parameter (α), a minimum biomass threshold may be required in addition to an F-based BRP. If environmental variability affects the density-dependent parameter (β), a biomass threshold may not be necessary. The effects of environmental variability make a difference in the way the population will respond to harvesting.

The advantages and disadvantages of fishing mortality rate and biomass-based reference points were discussed at length by the panel that examined the definitions of overfishing in U.S. fishery management plans (Rosenberg et al., 1994). Fishing mortality reference points are generally easier to calculate and more robust than biomass reference points when limited data are available. Biomass thresholds are needed, particularly for stock rebuilding, but there are still no widely accepted methods available for deriving biomass thresholds. Biomass thresholds can increase the average yield, with a corresponding increase in the variability of yield. Rosenberg et al. (1994) recommended that control laws combine a maximum fishing mortality rate and a precautionary biomass level wherever possible.

CLOSED-LOOP MANAGEMENT STRATEGIES

Harvest strategies based on fishing mortality rate or biomass targets modify harvest tactics depending on the stock assessment results and thus are "closed-loop" strategies in the terminology of control systems theory. In this closed loop, there are tight linkages: regulatory performance depends on assessment performance and assessment performance in turn depends on the regulatory pattern. Sustaining fish populations requires management procedures and practices that can cope with considerable uncertainty about population dynamics and the accuracy of stock assessments.

Input Control

Input controls limit inputs such as effort (labor) and capital that can be devoted to a fishery. Early research on management under uncertainty focused primarily on designing policies for dealing with unpredictable variation in recruitment and survival rates. Most of this research did not consider errors in assessments of stock size. A basic finding during the mid-1970s and later (Reed, 1974; Clark, 1985; Mangel, 1985; Hilborn and Walters, 1992) was that the optimum management in the presence of recruitment variation involves following a "feedback strategy" rule that specifies the best fishing rate and/or quota to select each year as a function of the stock size that year. The notion of harvest strategies has since become associated with such rules, in applications ranging from whales to salmon management to current National Marine Fisheries Service (NMFS) rules for varying fishing mortality rate with stock size.

Optimal rules were found generally to be simple, involving either (1) trying to maintain a fixed escapement[*]

---

[*]Escapement is the number of spawners remaining after harvest.

or base stock (*rule: harvest annual excess of biomass over base*) or (2) trying to maintain a fixed exploitation rate (*rule: harvest same proportion of biomass each year*). The shift from input control to output control (see below) resulted from the realization that determining the correct F or harvest is only part of the management problem. Implementing the proper F is also difficult. Implementation of output controls was motivated by a desire to implement F control as efficiently as possible, in an economic sense.

Output Control

During the late 1970s, considerable research was conducted by economists regarding how to improve the economic performance of fisheries, and this work led to the idea of "output control" via individual fishing quotas.[*] It was noted that ownership of predictable shares of allowable harvest can lead to much more efficient harvest planning and participation in management than occur in historical input control situations in which limits on effort or total quota lead to intense competition among fishermen for available fishing opportunities. Output and quota management policies have become increasingly popular with fishery managers, both because they are presumed to lead to better economic performance and because such policies simplify the manager's responsibilities. Under individual quota management, issues of catch allocation among stakeholders can be resolved by trading among stakeholders of their clearly denominated shares of the catch, with little intervention from managers. Responsibility for determining how large the total allowable catch (TAC) should be, given stock assessments and uncertainties, is often assigned by the manager to scientific assessment teams and processes.

There have been a few recent attempts to evaluate whether constant quota policies (*rule: take the same quota each year independent of stock size variations*) might still be economically optimum even if the quota were adjusted downward to allow for natural variation in recruitment (Hannesson, 1989; Arnason, 1993). The general result of these evaluations has been to recommend that quotas should not be held constant because in most cases a safe constant quota level would be too low compared to the long-term average MSY that could be obtained under a regime of variable annual catches. This has led to the recommendation that annual quota should vary with stock size and that quota shares should be set as proportions of the annual TAC.

Implementation Problems with Constant Exploitation Rate Strategies

Even where quota management has not proceeded to the individual fishing quota approach recommended by many economists, there has been wide acceptance of the simple strategy rule TAC = $F \times$ EB, where F is the target fishing mortality rate (e.g., F_{max}, $F_{0.1}$, $F_{X\%}$) and EB is the exploitable biomass. Constant exploitation rate or constant-F strategies are in principle very simple and have been shown to be nearly optimal in cases where the primary management objective is risk averse (Deriso, 1985) and in situations where long-term changes in the carrying capacity of marine ecosystems cannot be anticipated significantly in advance (Walters and Parma, 1996). These policies are robust to underestimation of the optimum F; using an F value lower than optimal for a given stock, although it will result in lower catches, can actually lead to higher biomass than optimum, and this increase in biomass partially balances the effect of using a lower-than-optimum F because the spawning stock is overprotected and gains will be achieved in later years. A variation on constant-F policies is to set a lower biomass limit, below which the target F is reduced to promote more rapid stock rebuilding (e.g., Quinn et al., 1990). With a depensatory stock-production relationship,[†] the optimal harvest strategy is to vary F in relation to stock size (Spencer, 1997).

Early research on performance of constant-F strategies presumed either that stock size is estimated accurately

---

[*]Another National Research Council study is being conducted at the request of Congress to evaluate the advantages and disadvantages of individual fishing quotas and other mechanisms for managing U.S. marine fisheries. The present study touches on quotas only as they are relevant to stock assessments.

[†]A depensatory stock-production relationship is defined by stocks in which per capita recruitment is reduced at low stock sizes.

or that F can be regulated directly by regulating fishing effort. Indeed, TACs and effort control were the two most commonly used tactics for implementing constant fishing mortalities. It was assumed in early assessment methods that $F = q \times$ Effort, with the catchability q representing a constant proportion of stock caught by one unit of effort. Variation in q over time was expected because of changes in fishing technology; therefore, many methods were developed for standardizing effort to some constant q value. However, a number of recent stock collapses, along with analyses of the behavior of fishers and fish, have indicated it is highly likely that q increases rapidly as stock size declines. That is, fixing effort is no guarantee of a safe F value; F can increase substantially as stock size declines if the geographic range used by the fish shrinks and fishers are able to track this contraction so as to target remaining fish concentrations efficiently, thereby increasing q. Such a situation has occurred, for example, with the Peruvian *anchoveta*.

Conversely, setting the TAC equal to the target F multiplied by an estimate of biomass is controversial because it has placed a significant responsibility on stock assessment to provide good estimates of biomass (or at least estimates that are not often greater than actual biomass). Simulation tests of long-term management in which TAC is set as a constant fraction of an uncertain biomass estimate indicate that biomass estimation errors with coefficients of variation (CV) greater than 50% are likely to result in substantial degradation in long-term harvest performance, particularly for highly productive stocks (Eggers, 1993; Frederick and Peterman, 1995; Walters and Parma, 1996). Following an approach similar to that described in Walters and Parma (1996), the committee simulated long-term performance of constant-F policies for two prototypical cases, referred to as "herring" (a short-lived, rather productive stock with optimal harvest rate equal to 0.25) and "cod" (a long-lived, less productive stock with optimal harvest rate equal to 0.10). The simulations were conducted by assuming that errors in the annual estimates of biomass were lognormally distributed and that population carrying capacity changed randomly from year to year with correlation $r = 0.3$ or 0.8 between total catch and the CV of biomass estimates (Figure 4.6). Three different harvest strategies were simulated: optimal escapement, optimal F, and risk-adjusted F.

Simulation results presented in Figure 4.6 indicate that degradation is more severe when estimation errors are strongly autocorrelated, for example, as expected when the estimates are based on fitting times series of data to population dynamic models. As expected, an increase in the CV of estimated biomass resulted in a lower catch for all harvest strategies and both levels of autocorrelation. However, the risk-adjusted F harvest strategy seemed to be less affected by increasing CVs. In addition, with autocorrelation, the probability that the stock falls to low, potentially dangerous levels may increase because successive errors no longer cancel out as they do to some extent when estimation errors are independent from year to year. This degradation could become critical in situations in which biomass estimates tend to be greater than actual biomass levels during stock declines. Such a situation typically occurs when catch-at-age analysis is tuned with CPUE data and CPUE does not decrease in proportion to abundance because fishers are targeting constricting stocks (this is illustrated by data sets 1-4 in the committee's simulations (see Chapter 5).

Faced with the possibility that inaccurate stock assessments will cause TAC (and hence F) to be set too high, the fishery manager has three strategic options:

1. ignore the danger and accept the degradation in long-term fishery performance resulting from the F variation (this degradation can be as much as 50% of potential fishery value [Walters and Parma, 1996]);
2. use a risk-adjusted biomass assessment in the TAC = $F \times$ EB calculation; or
3. use a risk-adjusted F assessment in the TAC calculation.

Simulation studies (Frederick and Peterman, 1995; Walters and Pearse, 1996) indicate that risk adjustment factors (i.e., the factor by which F is reduced to account for errors in the estimate of EB) may be substantial, approaching or exceeding $\exp\{-0.5\ CV^2\}$. In practice, risk adjustment factors do not have to be calculated, provided the harvesting strategies are evaluated by assuming realistic errors in the estimation of stock size. In the simulations shown in Figure 4.6, biomass estimates had a log normal error with a median of zero and bias equal to $\exp\{0.5\ CV^2\}$, so a risk adjustment equal to $\exp\{-0.5\ CV^2\}$ was used. The degradation in average performance of risk-adjusted F policies with increasing CV was much less severe than when an optimal equilibrium F policy was assumed. Yield losses could still be substantial for CV > 0.5 in some trials, especially when estimation errors were

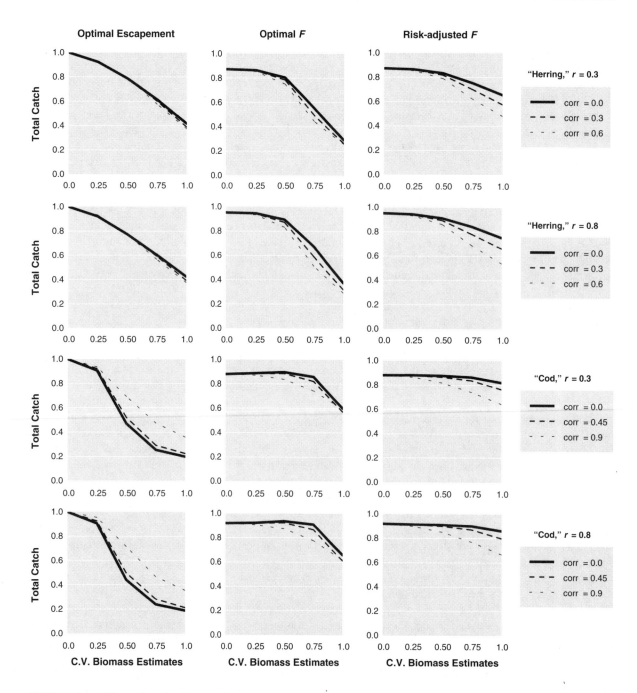

FIGURE 4.6   Effect of stochastic errors in annual biomass estimation on performance of escapement and fixed exploitation rate strategies. Average 100-year catches are expressed as a fraction of the maximum total yield achieved when all future recruitment deviations are known in advance and stock size is known without error. Optimal escapements are computed by using dynamic programming and assuming no estimation errors; optimal F corresponds to the fixed exploitation strategy that maximizes long-term average catches; risk-adjusted F = optimal $F \exp\{-0.5\ CV^2\}$, where CV is the coefficient of variation of biomass estimates. Three levels of autocorrelation in the estimation errors (corr) were used, with the maximum set equal to the growth-survival parameter of the delay-difference model used in the simulations (see Walters and Parma, 1996, for further details). Carrying capacity was assumed to fluctuate as an autoregressive random process with correlation given by r. Each plot is the average over 100 random series of environmental trends and 100 biomass estimation trials done on each of them.

assumed to be serially correlated. The probability that the stock biomass fell below 20% of average during the first 20 years of simulation (Figure 4.7) could still be high even under the risk-adjusted harvest rate, especially for very productive stocks (the "herring" case).

In summary, the two most common tactics for implementing fixed harvest rate policies—effort control and quota control—imply that managers must either obtain accurate biomass estimates to set TACs or use input regulation of fishing effort. Either approach entails the risk that F will increase dangerously with accidental decreases in stock size. This apparent dichotomy does not in fact cover all options for achieving constant F (Walters and Parma, 1996; Walters and Pearse, 1996). There are a variety of other tactics for preventing high F besides effort regulation, such as large marine refuges and direct assessment of cumulative F using tagging during a given season. Complementing quota management with space and/or time harvesting restrictions that directly limit the proportion of fish exposed to harvest may prove a costly but effective way of placing a firm upper bound on F.

When evaluating such regulatory schemes, the costs of placing a large number of fishers in competition in a substantially reduced time period and geographic area must be weighed against the costs of following more standard harvest tactics under realistic assessments of risks. The results of simulation experiments such as those cited above are particularly worrisome because they indicate that yield losses due to biomass estimation error in systems where $TAC_t = F\hat{B}_t$ can exceed the gains in economic performance expected from individual quota management (from reduced competition and more efficient deployment of fishing activity). In short, it does not make much sense to move to a quota management system if the quota must be corrected downward so far as to wipe out the gains from having it in the first place. Alternative regulatory approaches involving direct control of harvesting risks through area-time closures must be considered experimental at this time.

Improve Management Strategies or Improve Assessments?

Application of risk-adjusted (based on F or biomass) reference points would immediately lead to reduced TAC and create an economic incentive for investment in improved data gathering and assessment procedures to reduce the CV of biomass estimates (Pearse and Walters, 1992; Walters and Pearse, 1996). The committee can offer no clear and general recommendation about where to direct this investment from among the three plausible options:

1. increase sample sizes in existing data gathering programs (surveys, age composition sampling, tagging);

2. initiate new sampling programs to provide auxiliary data such as relative abundance of pre-recruits, egg surveys, and direct estimates of F based on tagging; and/or

3. invest in development and implementation of "better" assessment models for making more effective use of existing data.

Monte Carlo simulations of performance improvements that can be achieved by type 1 investments (e.g., Walters and Collie, 1989) suggest that only limited opportunities are available for reducing CVs of biomass estimates solely by gathering more of the usual data. In general, variability among observations in surveys and variability in age composition due to multinomial sampling appear to greatly underestimate the actual uncertainty in these key types of assessment data. That is, year-to-year and among-age variances in the measurements are substantially greater than expected from within-sample variance and sample size calculations, indicating that some major sources of variation are not accounted for in assessing the precision of these data sources. For example, survey abundance estimates often vary more from year to year than would be expected based on standard calculations of sampling variance (as in Chapter 2) and maximum variability that is biologically possible given recruitment and survival rates. Given this consideration, any assessment of the value of gathering more of the same data would be highly suspect, and would most likely be overly optimistic.

The development and use of Stock Synthesis models and related multi-data-type procedures are generally viewed as investments of types 2 and 3. There has been little exploratory work using Monte Carlo simulation to ask "what if we had data of type X?" questions related to available assessment procedures; only recently have

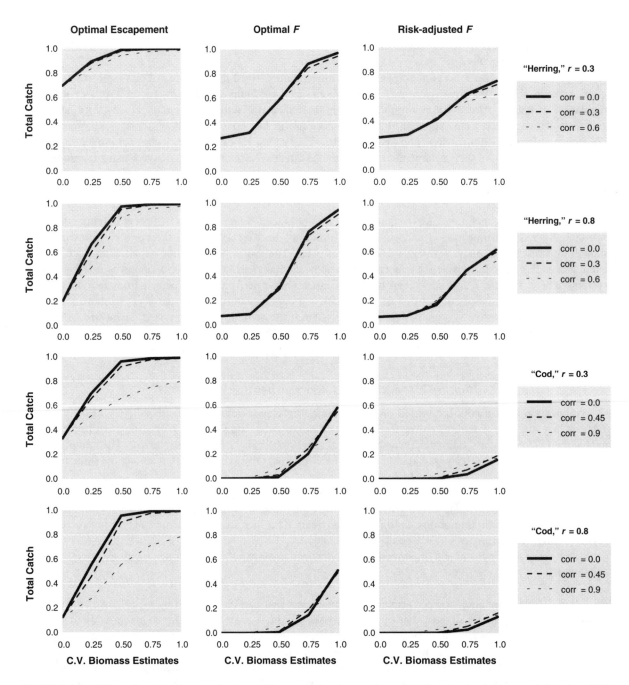

FIGURE 4.7   Effect of stochastic errors in annual biomass estimation on the probability that stock biomass falls below 20% of average carrying capacity at least once during the first 20 years of simulation. Each simulation starts with stock biomass at the optimal equilibrium size. Further details are provided in the caption for Figure 4.6.

multi-data assessment procedures such as Stock Synthesis (Methot, 1989, 1990, Bence et al., 1993) become available to allow realistic answers to such questions. It is not at all obvious that merely adding more auxiliary data and corresponding complexity in the assessment model (more "nuisance parameters," more structure) will produce more precise biomass estimates. Experience with simulation of closed-loop management performance using deliberately oversimplified assessment models has indicated that it can often be better (in terms of long-term average catch) to use a simpler assessment model that could be parameterized with available data (Hilborn, 1979; Ludwig and Walters, 1985, 1989). The basic problem here is that assessment using more complex, realistic models may provide more accurate estimates, but these estimates are likely to be less precise (have higher variance) than estimates from oversimplified models (Figure 4.8).

Basic statistical theory demonstrates the negative effects of overparameterization in cases where a sequence of increasingly complex assessment models is fit to a fixed set of historical data. It is not yet clear, however, how performance changes when there is simultaneous addition of model complexity *and* more auxiliary data. Although oversimplified models may outperform more realistic ones for implementing some closed-loop decision rules, the estimates of uncertainty about key management parameters (e.g., exploitable biomass) based on oversimplified models will be too optimistic and should not be used as a basis for policy evaluation. Thus, the optimal model for policy evaluation may be more complex than the optimal model for estimating population parameters in a stock assessment.

When Closed-Loop Management Strategies Are Most Likely to Fail

The two stock assessment situations most likely to result in persistent bias in stock size estimates occur when (1) stock size is collapsing rapidly and (2) stock size is very low and there is uncertainty about whether recovery is possible.

Stock Collapse

During stock collapses, overestimation of remaining abundance will lead to excessive harvesting, in which exploitation rate may even increase as the collapse proceeds (when quotas or TACs are not being reduced as rapidly as stock size). Overestimation of stock size during collapses has occurred primarily in situations in which (1) age-structured assessment methods have incorporated fishery CPUE data that do not decrease in proportion to stock size (cod stocks in Canada and herring in the North Sea are major examples; see simulated data set 1 in Chapter 5); and/or (2) where vulnerability of younger age classes has increased over time due to growth changes or changed targeting practices by the fishery (leading in virtual population analyses to an appearance that recruitment has been higher than actual).

Recovery Situations

During recoveries, overestimation of stock size will lead to overly optimistic assessments of how soon the fishery can be reopened or how much yield will be sustained in the future (e.g., northern cod stock in Canada from 1978 to 1985). Transient upturns in the indices of abundance may lead to false expectations of recovery when the political climate is favorable to relaxation of fishing restrictions (e.g., king crab from Kodiak, Alaska, in 1981; Orensanz et al., in press). Recovery trends following severe restrictions in fishing pressure are difficult to assess because catches are small relative to surplus production. Standard assessment methods do not work well unless fishery removals are significant because catches are generally the only quantities measured in absolute units that provide the scale of biomass estimates.

Fishery-independent surveys constitute the only way to guarantee that actual abundance trends will be detected by the assessment system, although having such surveys does not guarantee that trends will be assessed correctly, as some of the results of simulations reported in Chapter 5 indicate. Many surveys have been developed without careful consideration of how the spatial structure of stocks is likely to change over time, for example, surveys that are designed to cover areas of highest abundance or convenient sampling locations, rather than the full

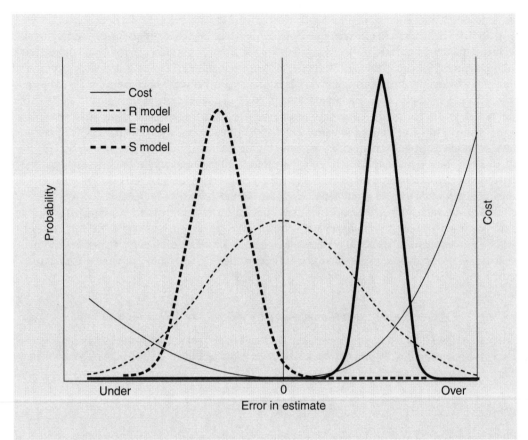

FIGURE 4.8   Uncertainty in estimates of policy parameters such as allowable catch depends on the model used for estimation. There is a cost to making estimation errors, which is generally asymmetric since overharvest tends to be more costly than
underharvest. Estimates can be quite variable (R case) when realistic models are used with uninformative data, resulting in
high expected costs (high odds of estimates being far enough off to produce high costs). Expected cost equals the sum of the
probability of an error multiplied by the cost at each potential error level. Estimates from very simple equilibrium models (E
case) fit to the same data may be quite precise but dangerously biased (i.e., the distribution is narrow but centered above zero).
Given existing limitations in modeling and sampling, which often preclude estimates that are both precise and accurate, the
most appropriate assessment model (S case) often is a deliberately oversimplified model that is not too badly biased (in a safe
direction), and gives less variable estimates than would be achieved with a more realistic model. SOURCE: Redrawn from
Walters (1996).

range of the stock. A survey with a poorly designed survey frame (too small a sampling area, near the "core" range
of the stock into which abundance contracts as the stock declines) may show the same incorrect trends as
commercial CPUE. For example, surveys of Pacific Ocean perch in Goose Island Gully off British Columbia did
not indicate a major stock decline in conjunction with high fishery removals during the 1960s; there is a nagging
fear among biologists that the stock was depleted far more severely than surveys indicated and that the survey
frame was too small (did not extend far enough into deep water) to detect the full extent of the decline.

RECREATIONAL MANAGEMENT STRATEGIES

Many coastal fisheries are changing rapidly from commercial to recreational status; large and effective
recreational fishing lobbies have had much political influence on allocation, basing their case on the higher

economic value of each fish taken by recreational fishing than by commercial fishing and on the fact that recreational anglers represent a large voting block (these forces are identified here to emphasize that the trend is not likely to be reversed). A number of serious assessment and harvest strategy problems result from this trend (see also Chapter 2), including the following:

• loss of abundance indices and catch composition data from the commercial fishery;
• drastic increase in the cost of basic catch and effort assessment because a larger number of more disperse units must be sampled;
• large changes in the space and time distribution of fishing and associated size or age vulnerability patterns; and
• entrenchment of an open-access system in recreational fisheries without any option for simple regulatory instruments such as total quotas or total effort limitation.

Biologists have not been too concerned in the past about overfishing in recreational fisheries, believing that recreational methods are generally relatively inefficient and that any depletion of stock should result in decreasing effort and cause a feedback reduction in harvest rate. This lack of concern about modern recreational fisheries is unwise; recreational methods are becoming much more efficient, recreational fishers are becoming more numerous, and they often target a mixture of species so that a decline in particular species is no guarantee that total effort and fishing mortality risk will decrease in response. Existing regulations, bag limits, and minimum sizes in recreational fisheries are unable to constrain recreational catches to target levels. Recreational harvests can greatly exceed target catches, sometimes by as much as 100%, so that total harvests are unsustainable even when commercial catch limitations are enforced. Because recreational fisheries are such a large component of fisheries harvests for many coastal fish species, it is necessary to include recreational fisheries in risk-averse harvesting strategies.

In principle, strategies used for varying exploitation rate with stock size in commercial fisheries could also be used in recreational fisheries to meet similar objectives for long-term sustained stock sizes and the opportunity for anglers to remove as many fish from the water as possible over the long term. In situations in which the recreational value of fisheries derives mainly from the ability to attract tourists to a region, the best management strategy might place more emphasis on maintaining high abundance of larger fish (and great attractiveness of the fishery compared to other options for the angler) than on high catch. However, such strategic trade-offs are not really the most important problem in recreational fisheries assessment; a significant difficulty has been in assessing the impact of alternative options for achieving desirable harvest rate strategies. Tactics for regulation of recreational fisheries include measures such as individual bag limits, size limits, space and time closures, and catch-release requirements. Evaluation of these measures almost always requires analysis of fish biology and fisher behavior at space-time-stock structure scales far more detailed than those used in commercial stock assessment. For example, a regulatory options model for Pacific salmon was used to examine chinook and coho stocks on time scales of weeks, to keep track of size-age structure in great detail, and to provide a complex accounting procedure for processes such as release of undersize fish and associated impacts on survival and later vulnerability to fishing (Argue et al., 1983). Such detailed calculations were necessary to even approach an accurate representation of the cumulative impact of such apparently simple policies as minimum size limits.

In general, there are few data on catch and effort from recreational fisheries. However, considerable information regarding age and growth, as well as other life-history characteristics is often available. Therefore, stock assessment scientists for recreational fisheries have tended to base their recommendations on yield-per-recruit and spawning biomass-per-recruit models. Some references dealing with these approaches include yield-per-recruit models for recreational fisheries (Buxton, 1992; Quinn and Szarzi, 1993), simulation models for minimum size limits and closed seasons (Atwood and Bennett, 1990), age and growth studies (Smale and Punt, 1991), and marine reserve evaluation by Atwood and Bennett (1995).

Not only are recreational fisheries more expensive to monitor and more difficult to regulate than commercial fisheries, they also may demand far more detailed and sophisticated assessment models for evaluating alternative harvest strategies. The development of such models is in its infancy, with some examples available in the literature

(Argue et al., 1983; Quinn and Szarzi, 1993; Szarzi et al., 1995). Conversely, relatively unsophisticated assessment models have been applied with success to recreational fisheries in South Africa (e.g., Smale and Punt, 1991; Buxton, 1992) and elsewhere.

IMPACT OF RESPONSIBLE FISHING ON MANAGEMENT STRATEGIES AND CLOSED-LOOP PERFORMANCE

The Food and Agriculture Organization adopted a global Code of Conduct for Responsible Fisheries on October 31, 1995, providing a framework for voluntary national and international efforts to exploit living aquatic resources in a sustainable manner while minimizing environmental impacts.* NMFS intends to adopt a version of this code after receiving public input. Bycatch is a major source of fish mortality (Alverson et al., 1994), and the move to responsible fishing practices may require major changes in fishing gear toward more selective methods to reduce bycatch. From a stock assessment perspective, movement to better gear types will add uncertainty about changes in age-size selectivity patterns, natural mortality, and pre-recruit mortality rates. Also, relative abundance indices based on historical fishery methods and statistics will be invalidated (this may actually be a benefit if CPUE statistics are as widely misleading as some scientists believe). For some fisheries, especially those in which assessment has been based on production modeling and relative abundance data rather than age-structured data, changing gear creates a situation similar to having no historical experience and starting assessments from scratch on a new fishery.

There are two main ways to guard against severe stock assessment problems during the transition to responsible fishing practices: (1) ensure that fishery-independent survey systems are in place for as long as possible before the transition begins, even if this means delaying the transition by a few years to gather baseline data; and (2) include regulatory measures (e.g., spatial closures) that directly limit exploitation rate during the transition period. The second of these points is particularly critical. It is often assumed that more selective and responsible fishing methods will be less efficient than the methods they replace; otherwise, the selective methods would already have been implemented for obvious economic reasons. This is not a valid assumption—many existing practices are maintained because the cost and risks of gear change are substantial, because regulatory requirements protect the existing gear type, and because the nonselectivity of existing gear leads to risk spreading by providing a mixed-species harvest. For example, in British Columbia there are proposals to replace river-mouth gillnet fisheries for Pacific salmon with traps and fish wheels. The latter gear types can be operated very selectively (live release of nontarget fish) but were banned in the past (at least in part) because they were believed to be too efficient rather than because they were not economical to operate.

EVALUATION OF MANAGEMENT PROCEDURES AND DECISION TABLES

Simulating Management Procedures

The interdependence between the performance of regulatory decision rules and assessment performance implies that alternative decision rules need to be evaluated in conjunction with the assessment method(s) that will be used to implement them. Alternative management procedures can be evaluated with a number of performance criteria using Monte Carlo simulations of stock projections that encompass a wide range of possible stock responses consistent with historical experience.

The first and probably most difficult step in developing a management procedure consists of identifying the range of possible stock responses to be included in simulation trials, either (1) as a discrete set of alternative "operating models" (Linhart and Zucchini, 1986), preferably with a probability assigned to each, or (2) as a Bayesian posterior probability distribution of model parameters from which operating models will be drawn at random. Unfortunately, usual criteria for model selection based on parsimony† are not directly applicable,

---

*This document was found at http://www.fao.org, June 25, 1997.

†A parsimonious model is the simplest model that can explain a given set of data.

because more complex models certainly will be rejected due to the inadequacy of the data (lack of statistical power), even when they may represent an equally plausible a priori hypothesis. The question of how much model uncertainty can be tolerated is an open one; some guidelines are beginning to emerge as more experience is gained with this type of exercise (Anon., 1995a).

The second step in evaluating alternative management procedures is to implement them in simulation trials. These trials should be designed in a way that yields realistic variability of model outputs. This step, although simple in theory, may require extensive computer time because realistic data collection schemes usually are not well described by simple probability distributions (e.g., Pelletier and Gros, 1991) and implementation of the assessment model may be very computationally intensive. Some simplifications may thus be needed. One such simplification is to abandon attempts to mimic the complex sampling schemes followed in practice and instead simulate data using standard sampling probability distributions, but with sample sizes much smaller than those used in real assessments, as an alternative means to achieve realistic sampling variances.

The second, more drastic, simplification commonly used in policy evaluation is to replace testing of the actual assessment method on simulated data by a much simpler simulation of the estimation errors that would affect the management parameters used in the decision rule. This simplification may be practical for some simple decision rules such as $TAC_t = F \hat{B}_t$, as long as (1) the errors affecting the estimates (the difference between estimated biomass and the "true" biomass) can be realistically characterized, and (2) assessment performance is relatively insensitive to the value of the control parameters (F in this case). Condition 2 is needed because the approach implies uncoupling the evaluations of performance of the decision rule and the assessment model. Pathological situations such as those simulated in data sets 1-3 (Chapter 5), in which the magnitude of the estimation errors changes as the stock is depleted, will not satisfy these requirements.

In the absence of such pathologies, there are still a few frequently disregarded points that are worth emphasizing. First, it is insufficient to characterize the estimation error affecting a single biomass estimate; performance of the decision rule will depend on the joint distribution of all future parameter estimates, so that quantification of serial correlation between estimation errors is important. Second, estimates of uncertainty derived from the usually oversimplified assessment models used to implement decision rules are not reliable; they are conditioned on oversimplified assumptions that result in reduced statistical variance and so will be negatively biased. The assumptions of error-free catch observations made in virtual population analyses and time-invariant selectivity of separable catch-at-age models are good examples. The estimated uncertainty about current stock size can increase substantially when the "separability" assumption is relaxed and selectivity parameters are allowed to change over time (Figure 4.9).

Similar observations apply to production model estimates; observation error estimators, generally advocated for providing more reliable point estimates of management parameters, result in gross underestimation of associated uncertainties in the presence of process error (Punt and Butterworth, 1993). The situation is more dramatic when model misspecification leads to pathological behavior of the estimates, which is evidenced by serious retrospective patterns but totally missed by standard estimates of variance derived using the same misspecified model (e.g., Parma, 1993). If the management procedure involves the use of an oversimplified model, then errors should be characterized by contrasting the estimates with the true quantities predicted or realized under the operating model. To assess serial correlations in the errors, the assessment model must be applied repeatedly to different segments of the same data set in much the same manner that retrospective analyses are conducted.

Decision Tables

Stock assessments can include an evaluation of the consequences of alternative management actions and are thus just a special case of a decision analysis. A decision analysis involves the following five steps (Walters, 1986):

1. Identify alternative hypotheses about the population dynamics or characteristics (often referred to as *states of nature*).

2. Determine the relative weight of evidence in support of each alternative hypothesis and express it as a

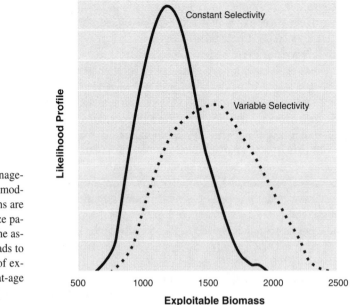

FIGURE 4.9 Estimates of uncertainty about management parameters derived from simple assessment models will be too optimistic if many of the assumptions are oversimplifications deliberately imposed to stabilize parameter estimates. Allowing for departures from the assumption of constant selectivity in this example leads to much wider probability profiles around estimates of exploitable biomass produced with a simple catch-at-age model (i.e., greater uncertainty).

relative probability. The weight of evidence is formally the Bayes posterior probability, which is increasingly being used in fisheries assessments (Walters, 1986; McAllister et al., 1994; Raftery et al., 1995; Punt and Hilborn, 1997).

   3. Identify each alternative management action.

   4. Evaluate the distribution and expected value of each performance measure, given the management actions and the hypotheses.

   5. Present the results to decisionmakers.

When there are a discrete number of alternative hypotheses and management actions, a "decision table" is an effective tool to summarize the decision analysis process and to present the results to decisionmakers. The example shown in Table 4.2 contains the following:

   • Alternative hypotheses about the true unexploited biomass level
   • Probabilities assigned to each hypothesis
   • Alternative management actions (TAC levels)
   • Consequences (in terms of some performance measure) of alternative actions (given in the body of the table)
   • Expected value of each management action (listed in right-hand column)

It is often possible to specify several different measures of population dynamics and stock condition as alternative hypotheses, and separate decision tables must be produced for each.

   Decision tables have the advantage that scientists and managers must explicitly examine alternative hypotheses about stock condition or population dynamics, the probability of these conditions, alternative actions, and their consequences. Decision tables are limited in the number of different states of nature that can be considered, however; in models with several uncertain parameters, all of the alternative conditions cannot be shown together, yet they must be integrated across stock conditions to produce summary statistics such as expected value and distribution of outcomes for each management action.

TABLE 4.2   A Simple Decision Table to Evaluate the Consequences of a Variety of Alternative Annual Total Allowable Catches (TACs)

	Alternative Hypothesis (unexploited biomass in thousand tons)						Expectation
	750	950	1150	1350	1550	1750	1046
Probability	0.099	0.465	0.317	0.096	0.020	0.003	
TAC[a] (thousand tons)							
100	0.51	0.63	0.70	0.75	0.78	0.81	0.66
150	0.26	0.45	0.56	0.63	0.69	0.72	0.49
200	0.22	0.26	0.42	0.52	0.59	0.64	0.34

[a]Shown in the interior cells of the table as ratio of stock size at end of management period to unexploited biomass.
SOURCE: Hilborn et al. (1994)

Biomass expectation = Σ Unexploited biomass$_i$ $\times$ Biomass probability$_i$
TAC expectation = Σ Unexploited biomass probability$_i$ $\times$ TAC probability$_i$

5

Simulations

SIMULATION APPROACH

The major goal of the committee's simulation study was to evaluate the performance of stock assessment methods and subsets of information (fishery, survey, ageing) for simulated fish populations where the true population parameters are known and where common assumptions usually made in stock assessments are violated. This project was similar in principle to a study by the International Council for Exploration of the Sea (ICES, 1993) Working Group on Fish Stock Assessment Methods, which compared a variety of age-structured methods. However, the violations considered herein are more severe than in the ICES study.

At its meeting on January 16-18, 1996, the committee designed the simulation model (the set of parameters and assumptions) that would generate simulated data sets to be used in the study. The committee used an age-structured model to generate 30 years of commercial catch and survey information as the basis of the simulations.* Complete details describing the procedure used to generate simulated data are given in Appendix E. The 30-year data series was longer than typically available because the committee was more interested in determining assessment failures due to violations of assumptions than in studying failures caused by shortness of the time series, although the latter problem also can be experienced in actual assessments. Simulated catch-age data were produced for ages 1-15, with the age 15 group containing information for all 15+ fish. The population was affected by natural mortality and fishing mortality; fishing mortality was an increasing function of age, as described by an asymptotic selectivity function. Fishing mortality also varied over time, with fishing effort being varied to achieve desired population trends and realistic variation. Recruitment to the population was governed by an asymptotic (Beverton-Holt) spawner-recruit relationship (Chapter 3) with a large, autocorrelated, environmental error component.

Certain features were included in the simulation model to test the robustness of stock assessment methods:

1. Ageing error: Many studies have shown that ageing error is a major problem in fisheries stock assessment (e.g., Summerfelt and Hall, 1987). Mean ageing error in the simulated data sets varied from 0 at age 1 to –1 year at age 15, with increasing variation as age increased. Ageing error was included because it corrupts information contained in the age composition data about year-class progression. To simplify the analyses, ages in the age 15+

---

*The simulated data and instructions will be available at the Ocean Studies Board site on the World Wide Web at http://www2.nas.edu/osb/.

group were not tracked; all fish in that group had the same probability of being misaged. This is not entirely realistic because older fish in this group are probably less likely to be aged outside the 15+ group (for example, at age 14). Nevertheless, the few fish in this group relative to other ages make this a minor concern (the effects of ageing error are discussed later in this chapter).

2. Fishery catchability changes: Fishery catchability was composed of two factors; one varied as a function of time and the other as a function of abundance. The first factor increased exponentially as time passed to mimic improvements in vessel efficiency due to technological improvements and learning. The second factor was a power function of abundance with an exponent of 0.4 (a little stronger than a square-root relationship). This factor was included to simulate the hyperstability* often observed in fishery catch per unit effort (CPUE). In a stock displaying hyperstability, CPUE tends to decrease more slowly than actual population size, leading to possible stock assessment errors and risk of population collapse because there are fewer fish available for harvest than indicated by CPUE trends (Hilborn and Walters, 1992).

3. Age selectivity† differences: For three of the five simulated data sets (1, 2, and 3), increased selectivity on younger fish by the fishery occurred in the last 10 years compared with the first 20 years, as shown in Figure 5.1. This feature was included because many assessments assume constant age selectivity and because selectivity changes can mimic changes in length-age relationships. Many actual fisheries appear to have changes in selectivity. For walleye pollock in the Bering Sea, changes in selectivity have resulted from changes in fishing patterns due to learning by fishers and spatial patchiness of fish populations (Quinn and Collie, 1990). When large year classes emerge, harvesters continue to target them as they age. For Pacific halibut, there is evidence of a substantial reduction in size of fish at a given age over the past 15 years (Clark, 1996). As a result, the selectivity of young age classes has been reduced because (1) the longline gear used is less efficient at catching smaller fish and (2) a larger fraction of the young individuals are below the legal size and must be discarded (Parma and Sullivan, 1996).

4. The age of 50% maturity was much higher than the age of 50% selectivity to the fishery (Figure 5.1). The model created a population that could be quite susceptible to overexploitation because fish reproduced at a greater age than the age at which they are recruited to the fishery. Such a situation is exemplified by cod, haddock, and flounder in U.S. waters, which start to be recruited at age 1 and mature at age 2+, although the difference is more pronounced in the simulated populations.

5. The survey gear had a dome-shaped selectivity function as shown in Figure 5.1 ("survey selectivity"). This choice was made because dome-shaped selectivity and natural mortality are often confounded in stock assessment applications. In addition, one data set (3) had doubled survey catchability for the last 15 years. This feature mimicked a change in survey vessel; analysts were told that a change of vessel occurred after 15 years.

6. Most stock assessment models assume constant natural mortality. In the committee's simulations, natural mortality was constant for fish of all ages during a given year but varied from year to year; this is probably true in actual populations due to variations in predation by other species and the changing incidence of disease with age. Natural mortality was modeled as a uniform random variable between 0.18 and 0.27 (with a mean value of 0.225).

7. In some fisheries, catch statistics are inaccurate, which most likely involves underreporting (see Chapter 2). One data set (2) included underreporting of catch by 30%.

8. Various process and measurement errors in the population's dynamics and the data were included in the model for realism, including random variation in recruitment, fishery catchability, survey catchability, fishery selectivity, fishing effort, ageing error, and sampling for ages.

Five data sets were generated with the age-structured model. The simulation model was constructed in an Excel spreadsheet. The simulation procedure can be visualized as

---

* "Hyperstability" is explained further by Hilborn and Walters (1992). Its counterpart is "hyperdepletion," in which CPUE decreases faster than the actual population. Hyperstability seems to be more common than hyperdepletion in actual populations.

† Age selectivity measures the vulnerability of different-age fish to the fishing gear relative to a reference age.

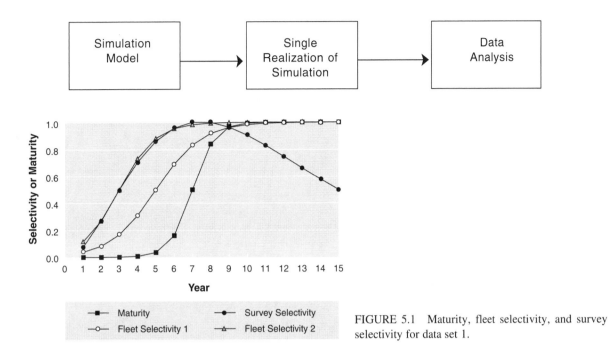

FIGURE 5.1 Maturity, fleet selectivity, and survey selectivity for data set 1.

The five simulated data sets differed in population trend, temporal changes in fishery selectivity, underreporting of catch, and changes in survey catchability (Table 5.1). The population trend simulated either a pristine stock being fished to a low level (data sets 1-4) or a depleted stock under recovery (data set 5), two common scenarios for fisheries. Fishery selectivity either was constant over time or declined stochastically from age 5 to age 3 in the last 15 years (see Appendix E). Although this design does not include all possible combinations of the four factors, the committee believed that the analysts could not have devoted the time to additional analyses. Quantification of these factors is described in Appendix E.

By comparing results from these data sets, the effects of particular factors can be understood. Data set 4 can be considered the easiest case; it has no changes in age-specific selectivity of the fishery or survey catchability over time and no underreporting. However, it did include changes in fishery catchability over time and as a function of biomass. Data set 3 can be considered the most difficult case because it includes changes in age selectivity as well as survey and fishery catchability. A comparison of results from data sets 1 and 2 shows the effect of underreporting. A comparison of results from data sets 1 and 3 shows the effect of the change in survey catchability. A comparison of results from data sets 1 and 4 shows the effect of the decrease in fishery selectivity. A comparison of results from data set 4 with data set 5 shows the effect of a decreasing population versus a recovering population.

The true exploitable biomass and the fishery and survey indices of exploitable biomass over time are shown for each data set in Figure 5.2. Each series has been scaled by its mean to show relative patterns over time. Except for data set 3, the survey index has the same pattern as biomass; for data set 3, the doubling of catchability causes the survey index to underestimate relative biomass at the beginning and to overestimate relative biomass at the end of the period. For each data set, the fishery index does not have the same trend as exploitable biomass because of the increasing catchability of the fishery over time, the decreased age selectivity in some data sets, and the dependence of catchability on biomass. The survey index is more variable than the fishery index; the survey relative error was 30% versus 20% for the fishery.

The committee sought assistance from National Marine Fisheries Service (NMFS) analysts who regularly use the major types of stock assessment methods for real assessments. The models tested (listed in order of complexity) included a production model; two versions of a delay-difference model; and age-structured analyses using ADAPT, a spreadsheet, Stock Synthesis, and Autodifferentiation Model Builder (ADMB, a commercial pack-

TABLE 5.1   Characteristics of Simulated Data Sets

Data Set	Population Trend	Age at 50% Selectivity	Underreporting	Survey Catchability
1	Depletion	Lower later	None	Constant
2	Depletion	Lower later	30%	Constant
3	Depletion	Lower later	None	Higher later
4	Depletion	Constant	None	Constant
5	Recovery	Constant	None	Constant

age).[*]   The spreadsheet implementation contained features similar to the Stock Synthesis program, but was a simpler implementation of the generic age-structured assessment (ASA) model. Further details about the implementation of these methods are given later in this chapter. The committee also utilized the services of a non-NMFS expert who performed additional ADAPT analyses. Data sets were sent to analysts in mid-March 1996 (see Appendix F for transmittal letter). Although the purpose of this exercise was to compare methods, the implementation of each method was affected by complex interactions among the individuals involved in the analyses, subjective and objective modeling decisions, the base model used, and the computer implementation of the model.

In addition to the five sets of catch, age composition, CPUE, and survey data from years 1 to 30, analysts were given growth and maturity parameters (see Appendix E), the ageing error probability matrix (Richards and Schnute, 1992), and information about the structure of the population and the data. Analysts were not provided with information about natural mortality, catchability, selectivity, the recruitment process, or the amount of underreporting (although they were warned that underreporting might have occurred). Analysts were requested to perform the analyses with three combinations of abundance indices:

1. CPUE data only (coded "F")
2. Survey data only ("S")
3. Both CPUE and survey data ("B")

Analysts were asked to perform these analyses independently, that is, not to use results of one analysis to initiate others or to work with analysts using other methods.

As mentioned earlier, the survey index has a greater relative error associated with it than does the fishery index. In other aspects, the survey index is less variable because surveys either are intentionally designed to be unchanging over time (e.g., same gear, sampling design, sampling methods, and sampling areas) or are changed only with associated calibration experiments to ensure that survey data are comparable over time. Thus, catchability and selectivity for the survey data were assumed to be constant (except for catchability in data set 3). Randomized sampling designs used in surveys reduce the hyperstability effect. Conversely, commercial fishers learn and change gear, fishing areas, and fishing methods to maintain or increase their harvests. The data sets provided to analysts are representative of differences known to exist between survey and fisheries data. The committee deliberately constructed the simulation according to conventional wisdom that a survey should provide a better index of abundance than data from a fishery. Nevertheless, the committee could have just as easily interchanged the fishery and the survey to make the survey the bad source of information. Hence, the simulation should be interpreted as having contained two indices of abundance, one of which was usually a good measure of abundance, and the other not.

Analysts' results were received near the beginning of May 1996 and summarized prior to the committee's

---

[*]Including ADMB in this analysis does not imply its endorsement by the committee. The model implemented using ADMB, however, possesses special characteristics not included in models commonly used by NMFS analysts. These characteristics include the ability to add a large number of process errors for recruitment, selectivity, catchability, and natural mortality. A Bayesian framework allows results to be synthesized in terms of posterior probability distributions for selected population parameters.

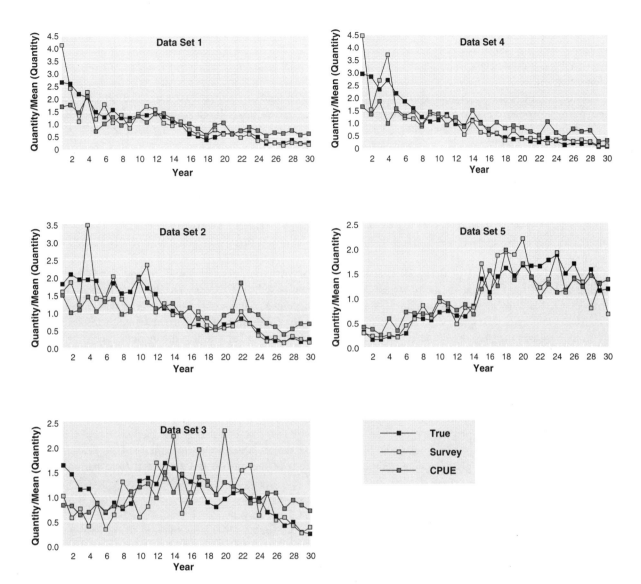

FIGURE 5.2   True exploitable biomass, survey biomass index, and fishery CPUE for the five simulated data sets. Data are plotted as quantity divided by the mean of the quantity over the 30 simulated years.

May meeting. Analysts cautioned the committee that the time they had available for the analyses was limited compared to a normal stock assessment. They noted that other information they would normally have available for doing a stock assessment (species characteristics, reports from harvesters and biologists, and other detailed information) was not given to them in this exercise. Such factors may have compromised the ability of analysts to obtain the absolutely best estimate, and the results presented herein do not necessarily reflect real-world conditions,[*] but all analysts operated under the same constraints, so this exercise constitutes a reasonable comparison of

---

[*]The simulation was not meant to replicate all the aspects of a real-world assessment, in which more time, human resources, and data about species would be available for an assessment and analysts would have greater freedom to reject bad data sets, conduct open discussion, and build models iteratively.

methods. One other caveat is that each data set represented the results of one replication of a stochastic process. The possibility that any given data set was extreme cannot be eliminated. Other caveats from the analysts will be discussed more completely in a planned National Oceanic and Atmospheric Administration (NOAA) Technical Memo that presents the analysts' reports.

The committee met with analysts on May 14, 1996. Each analyst presented the model results, as well as an indication of problems or insights gained in the analysis. The true values of biomass, recruitment, and exploitation were compared with analysts' estimates. The committee and analysts discussed what further work should be undertaken.

Three types of additional analyses were performed after the May meeting. First, because analysts estimated and used different values of natural mortality in the initial set of model runs (ranging from 0.15 to 0.25), the committee requested that they repeat the analyses with a common value for natural mortality equal to the true average natural mortality of 0.225. The committee asked four analysts to perform these in-depth analyses for the (1) ADAPT, (2) ASA spreadsheet, (3) Stock Synthesis, and (4) ADMB age-structured models. For each of the simulated scenarios, analysts were asked to use the three combinations of fishery and survey data, as in the previous analyses.

Second, a standard set of definitions of key management variables was agreed to and it was decided to calculate TAC (total allowable catch) in year 31 using a rate based on $F_{40\%}$ (as described in Chapter 4). Some analysts tried additional methods to improve the assessment; these results are reported later in this chapter.

Most existing assessment methods provide some estimate of precision of the parameter estimates. However, these are based on the structure of the assessment model being correct. Thus, unless the model structure is flexible enough to allow for major sources of uncertainty about the processes and data to be incorporated, estimates of precision tend to underestimate the true uncertainty in the assessment. It would have taken a considerable amount of additional work by both the analysts and the committee to evaluate estimates of uncertainty, so this was not done.

Finally, the committee decided to undertake retrospective analyses (see Chapter 3 for more detail about this subject) to determine the persistence of over- or underestimation of population parameters over time by the different methods. Although initial results indicated that the different methods were often able to recognize that the stock was severely depleted (or substantially recovered) at the end of the simulated time period, earlier recognition would be necessary for management to react in a timely fashion. Thus, retrospective analysis is important to determine how long it would take for the assessments to recognize underlying stock trends. The analysts were to use their methods on 16 subsets of the total data set: years 1-15, 1-16, . . . , 1-30.

The committee collected the results of these analyses in August 1996, summarized them, and conducted additional statistical analyses of the results to test whether trends of estimates and trends of the true values were parallel (even if biased). The committee met in August 1996 to review results obtained to that point, to formulate additional analyses of the simulation results, and to develop preliminary findings and recommendations.

SIMULATION RESULTS

This section describes the approaches taken for each specific model and presents the simulation results. The simulation results using mortality rates selected by the analysts are contrasted with results using the true M (natural mortality rate). Methods are compared on the basis of summary statistics related to important management parameters. Finally, the committee's additional analyses of model performance and retrospective analyses are presented.

Results Using Estimated M

Production Models

The production model used is described in Prager (1994). The production model estimates productivity parameters from total harvest and indices of biomass (CPUE, survey abundance over ages). No age-structured information was used. It was difficult for this model to produce reasonable estimates for many data set-data source combinations. In particular, no reasonable estimates could be obtained from data set 3, the most difficult one, because neither CPUE nor survey data were consistent indices of abundance over time.

Estimates by this method are shown in Table 5.2, along with the analyst's confidence in the results. Estimates were provided for MSY (maximum sustainable yield)[*] , F_{MSY}, E_{MSY}, B_{MSY}, B_{30}, and fishing mortality and biomass in year 30 relative to the MSY level (F_{rel} and B_{rel}, respectively).[†] The committee calculated relative fishing mortality from the fishing effort information and relative biomass from exploitable biomass values provided by the analyst.

The analyst's confidence in the results was generally low, with some moderate confidence in results from data set 4 (the easiest). When both fishery and survey data were used, estimates of MSY were 4 to 20% below the true value. However, estimates of absolute and relative fishing mortality and biomass generally were not close to the true values. Curiously, the use of fishery CPUE alone sometimes produced estimates close to the true values, even though the simulated CPUE was a biased measure of biomass. Nevertheless, the estimates often differed substantially from the true values, suggesting a lack of robustness of production models for data such as the simulated data used in this study.

The analyst cogently summarized the limitations of production models as follows: "The simulated data sets do not seem well suited to simple production modeling, and confidence in the quantitative validity of most of the results obtained is low. Noisy data, poorly correlated CPUE and survey indices, and relatively constant effort levels all probably contribute to this situation. Without knowing the underlying population model and conducting simulations, it is impossible to say to what degree age-structure effects also contribute. The apparently high fishing mortality rates and the extensive age-structured data available suggest that these fisheries are better suited to analysis by cohort-based methods. This suggestion is strengthened by several sets of simulated data in which apparently constant fishing effort leads to a population increase and then a decrease—such a scenario is incompatible with the assumptions underlying simple production models. This suggests either environmental forcing of recruitment, nonconstant catchability, or both."

Delay-Difference Models

Two analysts fitted the Schnute version of Deriso's delay-difference model (Deriso, 1980; Schnute, 1985), using total harvest data, the two indices of abundance, and a recruitment index (age 5 fishery or survey data for the first delay-difference method [DD] and age 3 [data sets 1-4] or age 4 [data set 5] fishery or survey data for the second delay difference method [DDKF]). For both models, recruitment was assumed to be knife-edge (i.e., all fish are vulnerable to the fishery at the same age). The population parameters estimated from this model therefore may not be strictly comparable to the true parameters, which were calculated from the age-dependent selectivity function.

The DD method was a measurement error model. Results from this model included estimates of biomass, recruitment, and fishing mortality. Natural mortality values were the same as those used in the first Stock Synthesis method described below (0.251. 0.251, 0.169, 0.201, and 0.191 for data sets 1 to 5, respectively).

For many of the DD results (Figure 5.3), deviations in exploitable biomass from the true value showed little trend, but there are substantial deviations in absolute values over the entire time period. Not surprisingly, the use of only fishery data (left panel of Figure 5.3) produced the most extreme deviations because the fishery data had a time trend in catchability that was not incorporated in the delay-difference models. Use of fishery data tended to lead to overestimates of biomass and failure to estimate the correct trend in exploitation fraction. Better results were obtained by using survey data alone or by using both data sources, but substantial discrepancies still remained. The deviations for data set 4 were particularly surprising, because this data set was relatively well

---

[*]MSY is given in metric tons (tonnes, t).

[†]Throughout this chapter, F = fishing mortality rate, E = effort, and B = biomass.

TABLE 5.2   Comparison of Results of Single Runs of Production Models with True Values[a]

| Data Set | 1 | 1 | 1 | True | 2 | 2 | True | 3 | True |
Index	F[b]	S	B_1		F	B_1		All	
MSY	580	90	300	312	995	218	240		315
F_{rel}	1.6	13	2.2	1.4	1.9	2.7	1.8		2.0
B_{rel}	0.18	0.11	0.23	0.14	0.07	0.21	0.19		0.36
Confidence	L[c]	L	L to N		L to N	L		N	
F_{MSY}	0.21	0.03	0.31	0.196	0.21	0.25	0.158		0.151
E_{MSY}	983	na[d]	1236	1223	917	1169	949		1096
B_{MSY}	2744	3183	954	1924	4639	1162	1820		2480
E_{30}				1703			1694		2139
B_{30}	480	430	236	276	324	322	346		903

| Data Set | 4 | 4 | 4 | 4 | True | 5 | True |
Index	F	S	B_1	B_2		F	
MSY	2470	300	430	480	513	635	564
F_{rel}	1.6	3.4	1.7	1.5	1.4	0.51	0.6
B_{rel}	0.03	0.11	0.17	0.18	0.05	1.4	2.09
Confidence	VL	L to M	L to M	L to M		L	
F_{MSY}	0.20	0.12	0.28	0.28	0.252	0.13	0.28
E_{MSY}	128	na	180	169	1827	2123	2025
B_{MSY}	12600	2540	1546	1701	2477	4738	2473
E_{30}					2552		1139
B_{30}	388	309	264	290	115	6950	5158

[a]The analyst examined all data sets and sources of data; those not given in this table were assigned no confidence by the analyst and productivity results were not meaningful. The analyst would not normally report absolute estimates F_{MSY} and B_{MSY}, believing that relative values are more accurate and more useful for management. They are included in this table for scientific interest. The major characteristics of the five data sets are given in Table 5.1   True values are those calculated by the committee from known parameter values.

[b]Data source: F = fishery; S = survey; B_1 = both, using standard methods; B_2 = both, using alternative techniques such as iterative reweighting or a combined survey-fishery index.

[c]Confidence: M = moderate, L = low, VL = very low, N = none

[d]Estimates of E_{MSY} are not available from production models using survey indices only. However, the estimate of F_{rel} (identical to E_{rel}) serves the same purpose and is often more useful in practice.

NOTE: F_{MSY}, E_{MSY}, and B_{MSY} are fishing mortality, effort, and biomass, respectively, at MSY. E_{30} and B_{30} are effort and biomass in year 30. F_{rel} and B_{rel} are fishing mortality and biomass in year 30 relative to F_{MSY} and B_{MSY}. E_{30} is a known value used to calculate F_{rel}.

structured. Nevertheless, the delay-difference model, by utilizing some information about age structure, showed improved estimates compared with production models. However, poor indices of abundance and the failure to use more age-structured data led to estimates that were more variable than the age-structured models discussed below. Models were rerun after the true average *M* value was provided, and those results are discussed in the following section.

ADAPT

The ADAPT approach is described in Gavaris (1988), Conser and Powers (1989), Restrepo and Powers (1991), and Conser (1993). It is essentially a cohort analysis on catch-age data where indices of abundance are included to estimate a relatively small number of parameters. The major assumption is that there are no errors in the catch-age data.

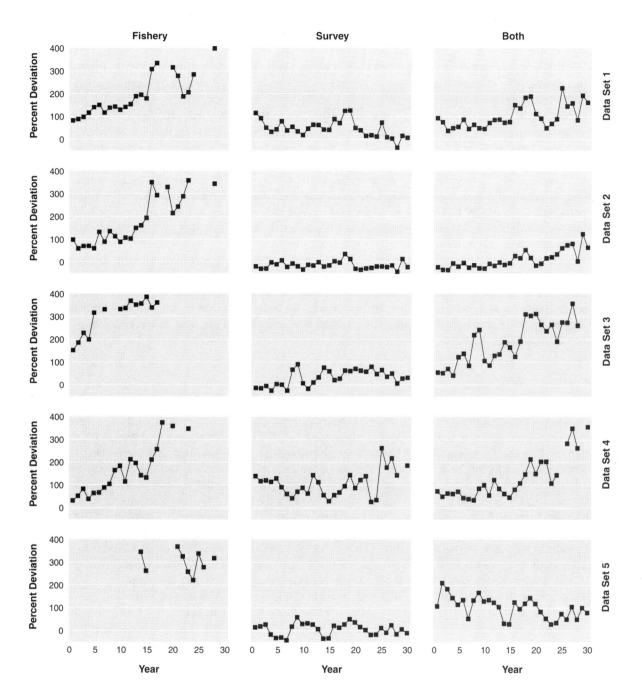

FIGURE 5.3 Percent relative deviations of estimated exploitable biomass from true exploitable biomass for model DD. Deviations greater than 400% are not shown in this figure and Figures 5.4 - 5.12. The major characteristics of the five data sets are given in Table 5.1.

The team of analysts who met for two days to perform the ADAPT analyses found that the ageing error matrix provided by the committee was more suitable to forward types of analyses such as ASA, SS, and ADMB and decided that they could not use such information. An M value of 0.15 was used in the ADAPT analyses, based on heuristic examination of the data and an attempt to estimate M from the data after making some assumptions about selectivity and catch variability.

Generally, the procedure used was (1) to examine the data and pool it beyond some age into a plus group, (2) to obtain selectivity estimates using a separable virtual population analysis, (3) to fix the ratio of fishing mortality for the plus group to the next-youngest age, and (4) to run a nonlinear least-squares procedure to estimate population parameters. Estimates of total and exploitable biomass, recruitment, spawning stock biomass, total biomass averaged over the year (B_{ave}), and average exploitation fraction (Y/B_{ave}, where Y = yield) were calculated. Age-specific estimates of catchability were examined and trends were discovered in some data sets. Therefore, except for data set 5, fishery CPUE data were not used, despite the committee's request for these sets of analyses, because the ADAPT team did not believe the information was useful. For data set 5, this group attempted a fit of the combined survey and fishery CPUE.

The relative trend in the estimates obtained was similar to the other age-structured methods (Figure 5.4). However, estimates of abundance were negatively biased, because the choice of natural mortality of 0.15 was too low compared to the true average M of 0.225. These results confirm the general conclusion that underestimating natural mortality leads to underestimation of abundance. A consequence of underestimating abundance is a tendency to overestimate exploitation rate. The team did not compute TAC because the procedure it would have used would have taken more time than was available.

Separable ASA Models

A family of age-structured assessment models is based on a statistical formulation of age-structured information and the assumption that fishing mortality is separable into age-selectivity multiplied by a full-recruitment fishing mortality (Doubleday, 1976; Fournier and Archibald, 1982; Deriso et al., 1985; Methot, 1989, 1990). A generic age-structured assessment model with these characteristics was formulated in a spreadsheet (ASA). It included a modification of the method assuming that all catch was taken halfway through the year and demonstrated that this approximation was accurate. The likelihood function consisted of a multinomial component that incorporated ageing error and a residual sum of squares term for the logarithm of each abundance index (number per boat-day from the fishery or survey catch in numbers, either pooled or by age). For data set 3, ASA assumed different catchabilities for years 1-15 and years 16-30 due to the change in survey vessel. An asymptotic curve for survey selectivity was used, but the analyst who ran ASA noted that he would have considered using a dome-shaped selectivity curve (the true situation) if he had other information to justify that choice. This spreadsheet model had similar characteristics to the Stock Synthesis model described next, but was simpler.

Estimates of natural mortality came from the Alverson-Carney procedure (based on longevity of a species and growth) and were 0.251, 0.251, 0.169, 0.201, and 0.191, for data sets 1 to 5, respectively. Trends from the survey and the fishery did not match, and the analyst suspected that the fishery data had trends in catchability. This analyst did a great deal of preliminary data exploration work to help him discover structural features in the data sets that improved his analyses. Deviations of estimated exploitable biomass from the true value were generally smaller when using only survey data or both compared to only fishery data (Figure 5.5). The ASA method usually resulted in less trend in the deviations over time than the methods previously described.

The Stock Synthesis method was implemented in a computer program as described in Methot (1989, 1990) and denoted SS-P in this report. A natural mortality value of 0.2 was used. Some models with constant recruitment were fitted to the data to mimic a production model. No results were presented from these analyses because recruitment was obviously not constant. The basic configuration of the model accounted for ageing error, used logistic curves for fleet and survey selectivity, and calculated fishing mortality values that equated the observed and predicted catch biomass values each year. The analyst calculated effective sample sizes from the mean squared error and showed that these were similar to actual sample sizes, which suggested that the SS-P

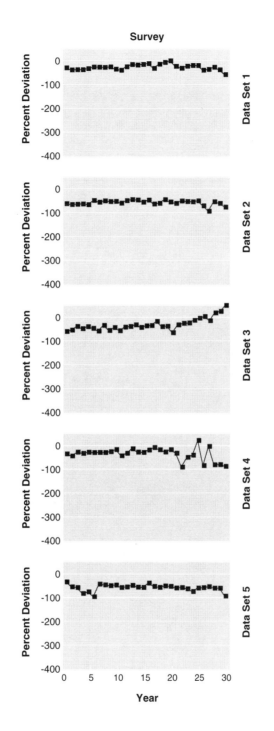

FIGURE 5.4   Percent relative deviations of estimated exploitable biomass from true exploitable biomass for the ADAPT model.

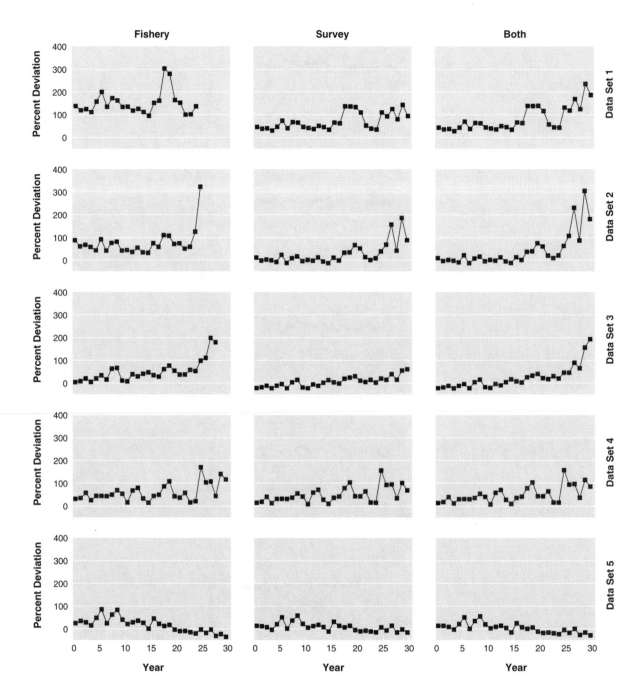

FIGURE 5.5   Percent relative deviations of estimated exploitable biomass from true exploitable biomass for model ASA.

models were not overparameterized. The program used was able to project population sizes into the future under a fishing mortality policy and could readily produce retrospective analyses.

AD Model Builder

This method was based on the work of Fournier and Archibald (1982) as incorporated into a computer software product called Autodifferentiation Model Builder. Analyses using ADMB were conducted by a team composed of one NMFS scientist and a colleague from outside NMFS. ADMB is similar in character to Stock Synthesis, with some additional model features. ADMB has the capability of adding process errors in the recruitment, selectivity, catchability, and natural mortality parameters. A Bayesian framework allows results to be synthesized in terms of posterior probability distributions for selected population parameters. Because the program code is compact, it was easy to carry out retrospective analyses. Of all the models, this method was the most complex, with some models having in excess of 400 parameters. Similar to other methods, a logistic curve was used for fishery and survey selectivities in the ADMB model.

The base model (ADMB1) assumed that fishing mortality at age and year was the product of age-specific fishery selectivity, catchability, fishing effort, and a process error term. Recruitment was assumed to be random about a mean value. The objective function was the sum of a modified χ^2 goodness-of-fit function of the age composition data, lognormal error terms for fishing effort and recruitment, and a lognormal prior distribution for M with a mean of 0.2 and a standard deviation of 0.15. In model ADMB2, additional parameters were added to the fishery selectivity function to account for changes in selectivity over time. As with previous models, results were better when survey data were used alone than when both data sets or fishery data alone were used (Figures 5.9 and 5.10). In model ADMB3, applied only to data set 1, additional parameters were added for natural mortality deviations over time.

Comparison of Exploitable and Total Biomass[*] by Data Set

Graphs of exploitable and total biomass for all age-structured methods conducted with estimated M values are shown in Appendix I for the five data sets. For data set 1 (Figures I.1 and I.6), most estimated series show the correct overall pattern of decline over time. The estimates of exploitable biomass are generally greater than the true biomass for the first part of the series, but most converge toward the true biomass near the end of the series. Notable exceptions are the models that use only the biased fishery data as an abundance index. It should be noted that overestimation observed in the early part of the series did not occurr with total biomass (Figure I.6).

For data set 2 (Figures I.2 and I.7), the pattern of decline was captured by models that used survey data. There was a tendency for the models to underestimate exploitable biomass. This difference from data set 1 can be attributed to the underreporting of catch that was included in data set 2.

For data set 3 (the "difficult" data set, Figures I.3 and I.8), models using only fishery data or those that did not use separate parameters for the two survey vessels tended to grossly overestimate exploitable and total biomass. Other deviations could have been caused by the incorrect specification of natural mortality and changes in age selectivity over time. All models overestimated biomass at the end of the series, sometimes dramatically. No model estimated exploitable biomass particularly well for this data set, although several models captured the overall trend.

For data set 4 (the "easy" data set, Figures I.4 and I.9), most models estimated the correct trend in biomass, showing that the population was depleted at the end of the series. As in other data sets, models using only fishery data tended to overestimate biomass, particularly at the beginning of the series.

For data set 5 (Figures I.5 and I.10), the only one with an increasing trend in biomass, all models showed an increase in exploitable and total biomass over time. However, there was a general tendency to underestimate the amount of recovery in the population. Unlike the other data sets, estimates were more accurate in an absolute sense at the beginning of the series than at the end.

---

[*]Total biomass is the sum of biomass of fish of all ages. Exploitable biomass is the sum of biomass multiplied by selectivity over age.

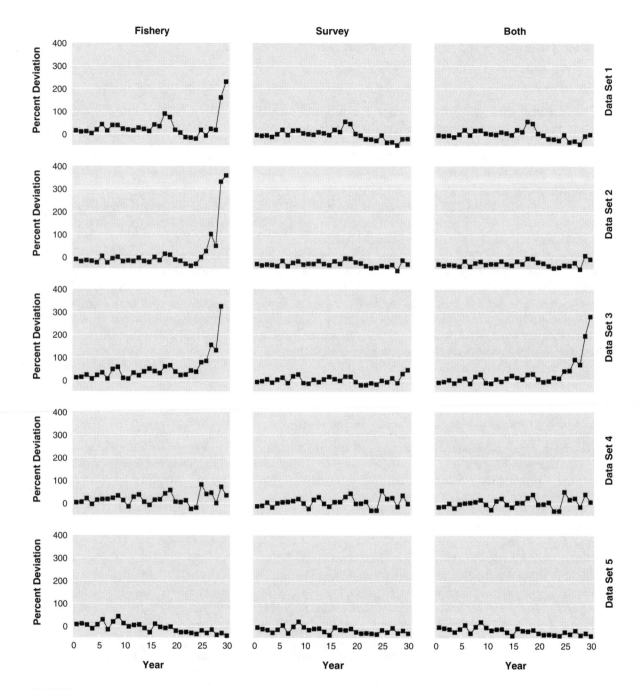

FIGURE 5.6   Percent relative deviations of estimated exploitable biomass from true exploitable biomass for model SS-P3.

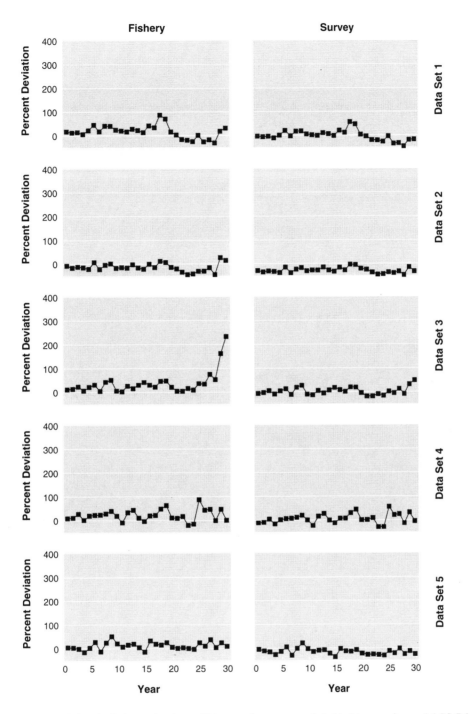

FIGURE 5.7 Percent relative deviations of estimated biomass from true exploitable biomass for model SS-P6.

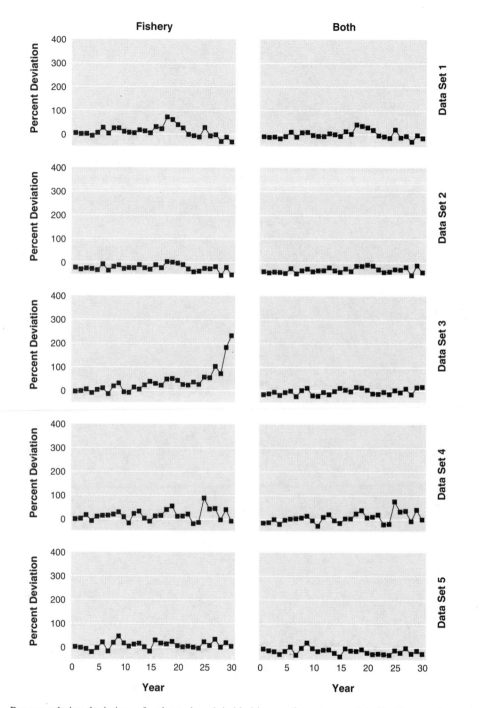

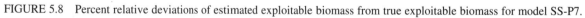

FIGURE 5.8   Percent relative deviations of estimated exploitable biomass from true exploitable biomass for model SS-P7.

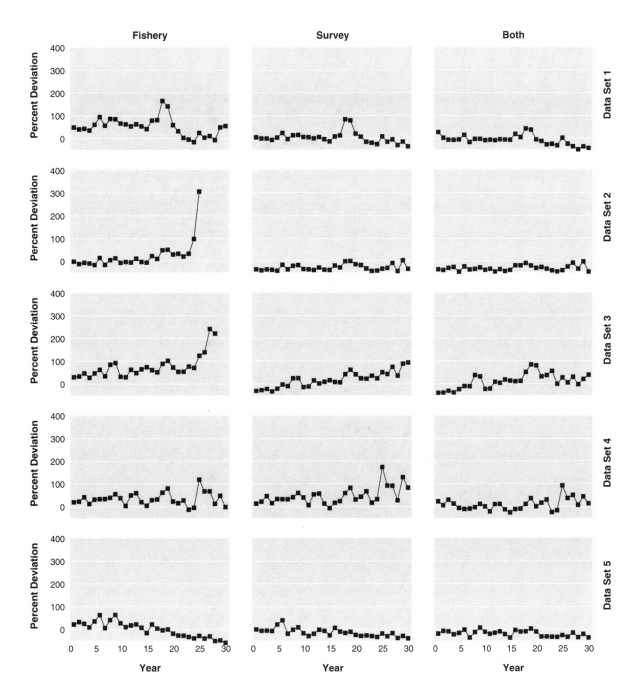

FIGURE 5.9   Percent relative deviations of estimated exploitable biomass from true exploitable biomass for model ADMB1.

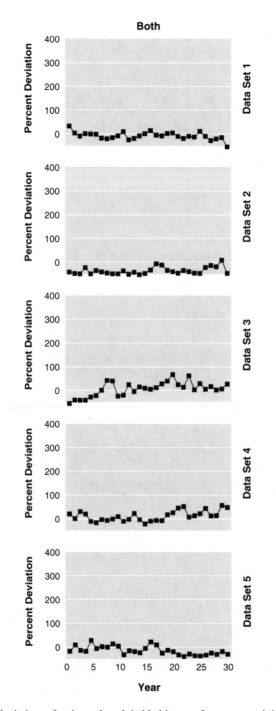

FIGURE 5.10   Percent relative deviations of estimated exploitable biomass from true exploitable biomass for model ADMB2.

Results Using True Average *M*

The delay-difference method DDKF and the four age-structured modeling methods were rerun using a true average *M* of 0.225. The following sections describe the results of the second set of model runs and discuss the effects of *M* value on stock assessment results.

Delay-Difference

The DDKF method included Kalman filtering and both measurement and process error. Results from DDKF included estimates of biomass and recruitment; true exploitation fraction was determined by dividing yield by biomass. Biomass and recruitment estimates were calculated using F and S data. For the process error model, the analyst conducted a separate investigation of the estimability of *M*. Results from DDKF were qualitatively similar to results from DD (Figure 5.11). One notable feature of the DDKF model was its ability to estimate accurately the amount of process error present in some data sources. This result suggests that stock assessment methods incorporating both measurement and process error reflect the uncertainty in both the data and the population more accurately.

ADAPT

The ADAPT method was rerun by an individual not associated with the group that performed the initial ADAPT analysis. The National Research Council's (NRC's) ADAPT method assumed (as was true) that a Gamma function describes the true shape of the gear selectivity of the simulated survey data sets and that a logistic selectivity curve describes the true shape of the gear selectivity of the simulated fishery data sets for the terminal year.

Separable ASA Models

Knowing the correct true average *M*, the analyst who ran the ASA model attempted to estimate natural mortality from data set 1 using only the survey index. The likelihood profile was centered about 0.28, which was further from the true value than was the original value used, suggesting that the available data did not provide useful information for estimating natural mortality. One possible explanation for this result is that the use of an asymptotic survey selectivity curve, when survey selectivity was actually dome shaped, resulted in confounding natural mortality with survey selectivity parameters (Thompson, 1994). For exploitable biomass in year 1, estimates were more accurate using the true average *M* for the first four data sets, but not for the fifth. For exploitable biomass in year 30, estimates were closer for data sets 1, 2, and 4, but not for 3 and 5. For TAC in year 31, estimates were closer to the truth for data sets 1, 2, and 5, but not for 3 and 4. Overall, estimates were more accurate using the true average *M* in 10 of the 15 cases, which is not significantly different from a 50:50 ratio (χ^2 test, $p = .59$).

Results from the SS-P3(S), SS-P7(F), and SS-P7(B) models with the true average *M* are shown in Table 5.3. For exploitable biomass, $F_{40\%}$, and TAC in year 31, overall estimates were more accurate using the true average *M* in 28 cases, less accurate in 15 cases, and the same in 2 cases; thus, the true *M* yielded significantly more accurate values for these three parameters (χ^2 test, $p < .05$).

ADMB

A final ADMB model (ADMB4) was constructed, using the true average *M* and incorporating time-series variations in fishery selectivity, fishery catchability, and natural mortality. For data set 3, the analyst chose not to incorporate two selectivity curves for the survey but rather allowed for time-specific deviations in catchability in the model and examined whether a trend in survey catchability occurred. Results are shown in Figure 5.12.

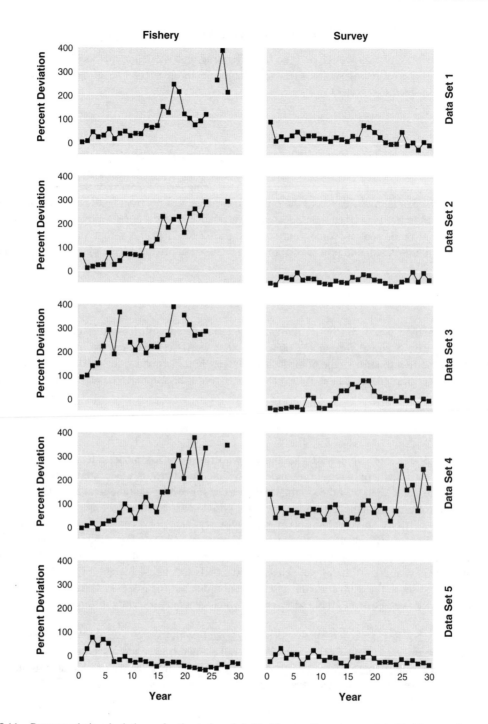

FIGURE 5.11   Percent relative deviations of estimated exploitable biomass from true exploitable biomass for model DDKF.

TABLE 5.3   Comparison of Some SS-P Results Using Original M (0.2) Versus True Average M (0.225)

Data Set 1	M	P3(S)	P7(B)	P7(F)
EB_{30}	0.2	220	234	118
EB_{30}	0.225	232	247	182
EB_{30} (true)		276	276	276
$F_{40\%}$	0.2	0.136	0.131	0.158
$F_{40\%}$	0.225	0.171	0.145	0.140
$F_{40\%}$ (true)		0.131	0.131	0.131
TAC	0.2	25	18	8
TAC	0.225	28	20	8
TAC (true)		23	23	23

Data Set 4	M	P3(S)	P7(B)	P7(F)
EB_{30}	0.2	113	114	105
EB_{30}	0.225	114	120	104
EB_{30} (true)		115	115	115
$F_{40\%}$	0.2	0.135	0.131	0.140
$F_{40\%}$	0.225	0.152	0.142	0.141
$F_{40\%}$ (true)		0.158	0.158	0.158
TAC	0.2	10	9	7
TAC	0.225	11	10	7
TAC (true)		13	13	13

Data Set 2				
EB_{30}	0.2	240	206	166
EB_{30}	0.225	251	223	157
EB_{30} (true)		346	346	346
$F_{40\%}$	0.2	0.118	0.116	0.132
$F_{40\%}$	0.225	0.141	0.126	0.123
$F_{40\%}$ (true)		0.119	0.119	0.119
TAC	0.2	27	15	9
TAC	0.225	30	18	8
TAC (true)		29	29	29

Data Set 5				
EB_{30}	0.2	3571	3788	5390
EB_{30}	0.225	4258	4576	5590
EB_{30} (true)		5158	5158	5158
$F_{40\%}$	0.2	0.159	0.151	0.160
$F_{40\%}$	0.225	0.173	0.172	0.163
$F_{40\%}$ (true)		0.182	0.182	0.182
TAC	0.2	447	439	627
TAC	0.225	527	565	681
TAC (true)		737	737	737

Data Set 3				
EB_{30}	0.2	1610	1512	2667
EB_{30}	0.225	1860	1829	2376
EB_{30} (true)		903	903	903
$F_{40\%}$	0.2	0.107	0.103	0.114
$F_{40\%}$	0.225	0.121	0.114	0.109
$F_{40\%}$ (true)		0.109	0.109	0.109
TAC	0.2	163	135	243
TAC	0.225	199	174	219
TAC (true)		70	70	70

NOTE: Results are given for three models: Model SS-P3 with only survey data, model SS-P7 with only fishery data, and model SS-P7 with both data sources. Estimates compared are exploitable biomass in year 30 (EB_{30}), the recommended fishing mortality rate ($F_{40\%}$), and the resultant TAC in year 31.

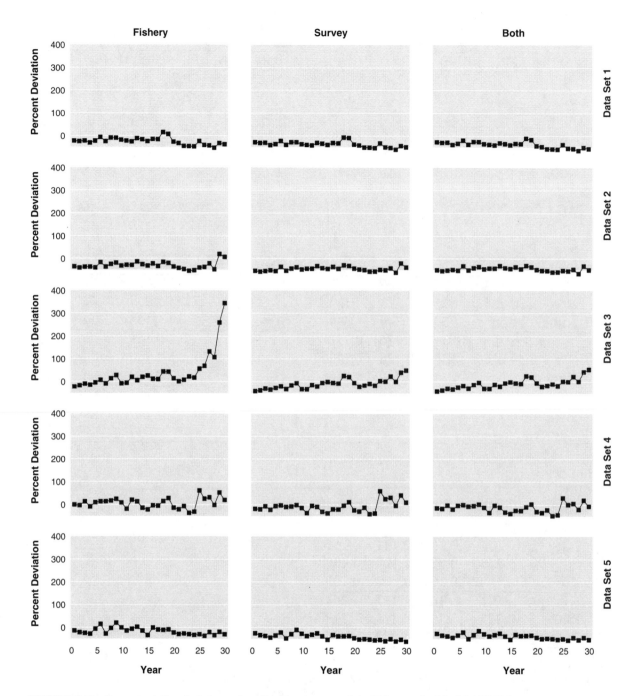

FIGURE 5.12   Percent relative deviations of predicted and true exploited biomass for Model ADMB4.

Comparison of Methods Using Summary Statistics

The statistics provided by stock assessments that are most useful for management include recent estimates of biomass, the amount of historical decline, recommended catch limits (TACs), recommended exploitation fractions, and spawner and recruitment information such as average spawning biomass and average recruitment. Specifically, the following information was summarized:

- Exploitable biomass in year 30 (EB_{30})
- Ratio of EB in year 30 to EB in year 1 (EB_{30}/EB_1), a measure of population decline or building
- Total allowable catch in year 31 from an $F_{40\%}$ exploitation strategy ($\widehat{TAC}_{31}$)
- Ratio of TAC_{31} to projected exploitable biomass in year 31 ($TAC_{31}/\widehat{EB}_{31}$), the projected exploitation rate. Projected exploitable biomass in year 31 was obtained by multiplying each analyst's estimate of exploitable biomass in year 30 by the true ratio of EBs in years 31 and 30. This was done to determine the recommended exploitation fraction ($TAC_{31}/\widehat{EB}_{31}$) in a consistent manner for all models, because some analysts did not provide estimates of EB in year 31.
  - Average recruitment ($\overline{R}$) from years 1 to 30
  - Average spawning biomass ($\overline{S}$)

Results for these management parameters are expressed in terms of relative error [(estimated - true)/true] for the various models, and results are shown separately for runs made with estimated versus the true M value in tables that follow. For the purpose of evaluating different models a modest goal for stock assessment is to obtain estimates of key management parameters within ±25% of the true values.

Results Averaged Across Data Sets

A summary of the results across data sets is presented to show the general performance of the models. The summary statistic used was the average of the absolute values of relative errors among the five data sets. The average of absolute relative errors is used to illustrate the precision of estimates compared to true values and the average of relative errors are used to illustrate the accuracy or bias. The absolute value indicates models that deviated in either direction from the truth (Table 5.4); results from individual data sets are presented in Tables 5.5 to 5.9 and can be used to examine the direction of the errors for the individual sets. Models that did not have results for all five data sets were excluded.

Results given in Table 5.4 show that there can be great deviations in management statistics from the true values. Figure 5.13 shows the results for exploitable biomass in year 30, with relative absolute deviations often exceeding 50%, comparing estimated M versus the true average M. Model runs that used only fishery data generally yielded the worst results, those that used only survey data were generally the best, and those that used both sources ranged from best to intermediate in their results. The success of runs with both data sources depended greatly on which model was used. Model runs with more complex treatment of fishery data (SS-P6, SS-P7, ADMB4) tended to have lower errors than those with simpler treatment (DD, ASA, SS-P3, NRC ADAPT), and a similar result was obtained when fishery data were used alone. Results obtained with the true average M were not much different from those with estimated M values, probably because the true average M used in the revised analyses was, in many cases, not that much different from the M used initially. The fact that the true population had varying M over time whereas analysts used constant M (except ADMB4) may have been more important than changes in average M.

Results for the ratio of biomass in year 30 to that in year 1 were similar to those for EB_{30} (compare Table 5.4B). This result was somewhat surprising because the committee expected it would be easier to estimate the overall change in populations than their absolute biomass levels.

The committee thought it was very important to consider the relative error in TAC, because this statistic is frequently the end point of assessment, being the recommended catch level to be taken in the next year.[*] The

---

[*]In the assessment trials, the maturity and growth schedules were known without error. In a real assessment, estimation errors in the maturity and growth schedules would contribute additional uncertainty to the estimated TAC values.

TABLE 5.4A   Average Absolute Values of Relative Errors [|(estimated − true)/true|] for Important Management Parameters[a]—Results with Estimated M

Model	Data source	EB_{30}	$\dfrac{EB_{30}}{EB_1}$	TAC_{31}	$\dfrac{TAC_{31}}{\widehat{EB}_{31}}$	$\bar{R}$	$\bar{S}$
SP	F	na	na	na	na	na	na
SP	S	na	na	na	na	na	na
SP	B	na	na	na	na	na	na
DD	B	2.24	1.18	na	na	na	na
DD	F	8.59	3.90	8.28	**0.19**	na	na
DD	S	0.51	0.30	0.42	**0.23**	na	na
ADMB1	B	0.36	0.47	na	na	**0.21**	0.28
ADMB1	F	11.49	11.25	na	na	0.90	0.29
ADMB1	S	0.58	0.65	na	na	**0.17**	0.29
ADMB2	B	0.41	0.59	na	na	**0.22**	0.28
ASA	B	1.35	1.28	2.09	0.27	**0.21**	**0.15**
ASA	F	17.82	8.82	39.77	0.73	1.37	0.50
ASA	S	0.66	0.57	0.95	**0.16**	**0.21**	**0.15**
SS-P3	B	0.67	0.84	1.20	0.36	**0.24**	**0.19**
SS-P3	F	2.25	2.15	3.94	0.46	**0.20**	**0.17**
SS-P3	S	0.26	**0.23**	0.32	0.26	**0.21**	**0.19**
SS-P6	B	0.26	**0.22**	0.37	0.28	**0.21**	**0.19**
SS-P6	F	0.56	0.50	1.11	0.37	**0.18**	**0.11**
SS-P7	B	**0.20**	**0.18**	0.32	**0.15**	**0.24**	**0.22**
SS-P7	F	0.66	0.66	0.99	0.33	**0.19**	**0.09**
ADAPT	S	0.68	0.93	na	na	0.54	0.44

NOTE: Values in table × 100 indicate percentage deviation of estimated values from true values. Boldface indicates averages within 25% of the true values.

[a]Values are summarized across data sets.

TABLE 5.4B   Average Absolute Values of Relative Errors [|(estimated − true)/true|] for Important Management Parameters[a]—Results with $M = 0.225$

Model	Data source	EB_{30}	$\dfrac{EB_{30}}{EB_1}$	TAC_{31}	$\dfrac{TAC_{31}}{\widehat{EB}_{31}}$	$\bar{R}$	$\bar{S}$
DDKF	F	5.62	3.89	8.49	0.42	0.36	na
DDKF	S	0.51	0.32	0.81	0.30	**0.22**	na
ADMB4	F	0.88	1.20	2.23	0.81	**0.18**	**0.14**
ADMB4	B	0.44	0.51	0.51	0.82	**0.13**	**0.20**
NRC ADAPT	F	3.45	4.01	8.25	1.46	0.29	0.80
NRC ADAPT	S	0.55	0.67	1.21	0.84	**0.24**	0.79
NRC ADAPT	B	0.95	0.89	1.57	0.75	0.26	0.82
ASA	S	0.66	0.47	1.09	0.28	**0.13**	**0.22**
SS-P3	B	0.82	0.78	1.53	0.38	**0.19**	**0.14**
SS-P3	F	2.44	1.86	4.46	0.47	**0.23**	**0.18**
SS-P3	S	**0.24**	**0.20**	0.37	0.29	**0.09**	**0.09**
SS-P6	B	**0.24**	**0.19**	0.36	0.30	**0.09**	**0.08**
SS-P6	F	0.57	0.34	1.11	0.39	**0.17**	**0.18**
SS-P7	B	**0.18**	**0.15**	0.27	**0.11**	**0.08**	**0.11**
SS-P7	F	0.52	0.47	0.78	0.32	**0.13**	**0.15**

NOTE: Values in table × 100 indicate percentage deviation of estimated values from true values. Boldface indicates averages within 25% of the true values.

[a]Values are summarized across data sets.

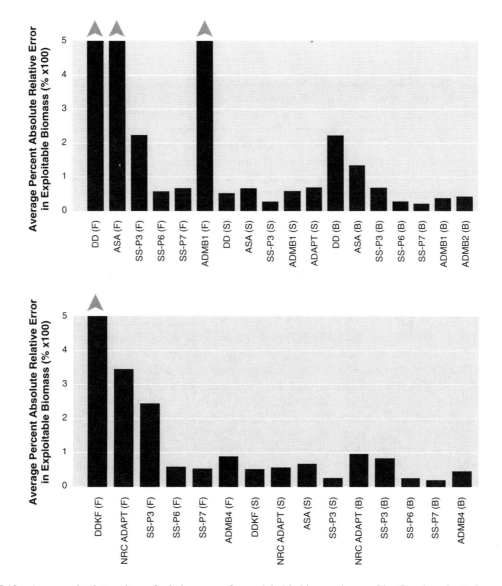

FIGURE 5.13   Average absolute values of relative errors for exploitable biomass in year 30.   (Results prior to [upper panel] and after [lower panel] knowing the true *M*.)

relative error in TAC was comparable to the relative error in exploitable biomass, sometimes higher and sometimes lower (see Table 5.4). In many cases, it deviated more than 50% from the true value. As before, models using only survey data or both data sources fared better than those using only fishery data. The ratio of TAC to exploitable biomass ($TAC_{31}/\hat{EB}_{31}$) is the recommended exploitation rate. The amount of relative error in this statistic was lower overall than for the previous statistics, especially for model runs using only fishery data (Table 5.4). Many model runs yielded relative errors well below 50%. For this statistic, the model used seemed also to be a factor: the NRC ADAPT and ADMB4 models had greater errors than other models, regardless of which combination of data was used. Average recruitment ($\overline{R}$) and average spawning biomass ($\overline{S}$) were estimated with less error overall than other management parameters. This is not surprising because these statistics are calculated over the entire time period, which averages out the positive and negative errors found in individual years.

Results by Data Set

Using Estimated M

Tables 5.5 to 5.9 present the relative errors in management variables for data sets 1 to 5, respectively. These tables show that average recruitment ($\overline{R}$) and average spawning biomass ($\overline{S}$) are estimated with less error overall than other management parameters. The ADAPT model generally underestimated these because the natural mortality used was too low. Models using only fishery data (F) often overestimated average biomass, unless the models contained sufficient additional parameters to overcome the catchability patterns. For data set 2, many models underestimated $\overline{R}$ and $\overline{S}$ because of underreporting. The same thing occurred with data set 5 because biomass and recruitment were underestimated for the recovering population. For the first four management parameters related to exploitable biomass and TAC, the amount of error is quite variable among parameters and among models. The worst results are generally obtained by using only fishery data for all data sets. However, the tendency for most models using survey data to underestimate biomass in data set 5 results in the fishery-only models sometimes producing estimates that are closer to the truth for this data set. The underestimation could have been due to the particular set of random numbers used for this data set, so it would not be appropriate to conclude that assessments based on commercial CPUE generally would be better.

For data set 1, examination of Table 5.5A shows that the goal of being within 25% of the true value was achieved for at least two of the first four management parameters (related to exploitable biomass and TAC for the DD(S), SS-P3(B), SS-P3(S), SS-P6(B), and SS-P7(B) models [notation indicates model followed by data set in parentheses]). In only one case (SS-P7[B]) however, was this goal obtained for all four management parameters. For data set 2, the 25% criterion was achieved for at least two parameters with the DD(S), SS-P3(S), SS-P6(B), SS-P6(F), and SS-P7(B) models, but no model met the criterion for all four parameters (Table 5.6A). For data set 3, the "hard" data set, the criterion was achieved for two parameters only for the DD(S) and SS-P7A(B) models, and neither of these models achieved the criterion for all four parameters (Table 5.7A). For data set 4, the so-called easy data set, the criterion was achieved for the DD(S), AD(F), AD(B), SS-P3(S), SS-P3(F), SS-P3(B), SS-P6(F), SS-P6(B), SS-P7(F), and SS-P7(B) models, but only SS-P3(S) achieved the criterion for all four parameters (Table 5.8A). For data set 5, the criterion was achieved for at least two parameters by the DD(F), DD(S), ASA(S), SS-P6(F), SS-P6(B), SS-P7(F), and SS-P7(B) models (Table 5.9A). The criterion was met for all four management parameters only for the ASA(S) and the SS-P7(F) models. The DD(S) model met the criterion for at least two parameters for all five data sets. Otherwise, no single model performed superlatively across all data sets and all management parameters.

Using True Average M

Results using the true average *M* value are presented in Tables 5.5B to 5.9B. By using the true average *M* value, $\overline{R}$ and $\overline{S}$ are estimated with even less error overall than results obtained without knowledge of the true average *M*. This probably results from a combination of use of the true average *M* and use of more complicated models in the second round of model runs that corrected some biases in the earlier models. However, in some cases, using the correct *M* in a model did not result in an improvement. For example, for data set 3, the second-run models tended to overestimate average recruitment, which did not occur with the first run. Whether this was due to the variability in *M*, confounding of model parameters, or random chance is unknown.

For data set 1, relative errors were less than 25% for at least two of the exploitable biomass and TAC statistics for the NRC ADAPT(S), SS-P3(S), SS-P3(B), SS-P6(B), SS-P6(F), and SS-P7(B) models; only the latter model met this criterion for all four parameters (Table 5.5B). For data set 2, at least two statistics achieved the criterion for the ADMB4(B), NRC ADAPT(S), SS-P3(S), SS-P3(B), SS-P6(F), SS-P6(B), and SS-P7(B) models. No model achieved the criterion for all four parameters, because the underreporting led to underestimation of biomass with resultant effects on TAC (Table 5.6B). For data set 3, the 25% criterion was achieved for two statistics only by the DDKF(S) model (Table 5.7B). As with the prior results, there was a general tendency to overestimate biomass and TAC statistics for this data set, often by a wide margin. For data set 4, all models (except DDKF and

TABLE 5.5A Data Set 1: Summary of Relative Error [(estimated – true)/true] in Parameters Important for Management—Results with Estimated *M*

Model	Data source	EB_{30}	$\dfrac{EB_{30}}{EB_1}$	TAC_{31}	$\dfrac{TAC_{31}}{\widehat{EB}_{31}}$	$\bar{R}$	$\bar{S}$
SP	F	0.74	na	7.17	0.74	na	na
SP	S	0.56	na	**-0.17**	0.56	na	na
SP	B	**-0.14**	na	4.39	**-0.14**	na	na
DD	B	1.61	0.35	na	na	-0.31	na
DD	F	6.04	2.88	7.96	0.27	na	na
DD	S	**0.06**	-0.51	**-0.09**	**-0.14**	na	na
ADMB1	B	-0.44	-0.56	na	na	-0.20	-0.17
ADMB1	F	0.54	**0.04**	na	na	-0.05	0.33
ADMB1	S	-0.36	-0.39	na	na	0.08	-0.26
ADMB2	B	-0.54	-0.66	na	na	-0.22	-0.17
ASA	B	1.86	1.00	2.96	0.38	0.38	0.29
ASA	F	45.27	18.42	101.35	1.21	3.29	1.19
ASA	S	0.94	0.33	1.35	**0.21**	0.32	0.29
SS-P3	B	**-0.01**	**0.03**	0.48	0.50	-0.11	-0.10
SS-P3	F	2.30	1.86	4.96	0.81	0.10	0.05
SS-P3	S	**-0.20**	**-0.17**	**0.09**	0.37	-0.13	-0.10
SS-P6	B	**-0.19**	**-0.15**	**0.09**	0.34	-0.13	-0.10
SS-P6	F	0.30	**0.12**	0.91	0.48	-0.09	0.05
SS-P7	B	**-0.15**	**-0.07**	**-0.22**	**-0.08**	-0.15	-0.14
SS-P7	F	-0.32	-0.36	-0.65	-0.49	-0.16	-0.02
ADMB3	B	-0.48	-0.57	na	na	-0.31	-0.23
ADAPT	S	-0.58	-0.40	na	na	-0.46	-0.36
TRUE		276	0.088	23	0.109	309	720

NOTE: For the first four management statistics, values within 25% of the truth are shown in boldface type.

TABLE 5.5B Data Set 1: Summary of Relative Error [(estimated – true)/true] in Parameters Important for Management—Results with Estimated *M*

Model	Data source	EB_{30}	$\dfrac{EB_{30}}{EB_1}$	TAC_{31}	$\dfrac{TAC_{31}}{\widehat{EB}_{31}}$	$\bar{R}$	$\bar{S}$
\DDKF	F	4.07	3.96	7.52	0.68	-0.56	na
DDKF	S	**-0.08**	-0.51	0.35	0.46	-0.55	na
ADMB4	F	-0.38	**-0.20**	0.39	1.24	-0.01	-0.03
ADMB4	B	-0.57	-0.41	**0.09**	1.51	-0.02	-0.10
NRC ADAPT	F	4.34	4.44	9.04	0.88	0.26	0.23
NRC ADAPT	S	**0.08**	**0.12**	0.35	**0.25**	0.04	0.21
NRC ADAPT	B	0.37	0.35	0.70	**0.24**	0.05	0.21
ASA	S	0.67	0.34	0.89	**0.13**	0.10	0.14
SS-P3	B	**0.04**	**-0.04**	0.68	0.61	0.03	0.01
SS-P3	F	2.18	1.41	4.98	0.88	0.25	0.17
SS-P3	S	**-0.16**	**-0.22**	**0.22**	0.45	0.01	-0.01
SS-P6	B	**-0.13**	**-0.20**	0.28	0.48	0.02	-0.01
SS-P6	F	**0.22**	**-0.07**	0.86	0.53	0.06	0.15
SS-P7	B	**-0.11**	**-0.13**	**-0.13**	**-0.03**	-0.01	-0.05
SS-P7	F	-0.34	-0.45	-0.67	-0.50	-0.02	0.08
TRUE		276	0.088	23	0.109	309	720

NOTE: For the first four management statistics, values within 25% of the truth are shown in boldface type.

TABLE 5.6A Data Set 2: Summary of Relative Error [(estimated − true)/true] in Important Management Parameters—Results With Estimated M Values

Model	Data source	EB_{30}	$\dfrac{EB_{30}}{EB_1}$	TAC_{31}	$\dfrac{TAC_{31}}{\widehat{EB}_{31}}$	$\bar{R}$	$\bar{S}$
SP	F	**−0.06**	na	3.66	3.96	na	na
SP	B	**−0.24**	na	1.79	na	na	na
DD	B	0.64	1.13	na	2.68	−0.40	na
DD	F	6.64	2.86	8.10	**0.19**	na	na
DD	S	**−0.22**	**−0.03**	−0.45	−0.29	na	na
ADMB1	B	−0.48	**−0.17**	na	na	−0.38	−0.60
ADMB1	F	48.85	49.84	na	na	3.42	0.06
ADMB1	S	−0.35	**0.02**	na	na	−0.35	−0.54
ADMB2	B	−0.47	**−0.08**	na	na	−0.40	−0.63
ASA	B	1.81	1.58	2.59	0.27	0.04	−0.08
ASA	F	36.33	18.84	83.66	1.27	3.01	0.53
ASA	S	0.89	0.70	1.21	**0.17**	−0.02	−0.09
SS-P3	B	**−0.08**	0.33	0.34	0.47	−0.36	−0.40
SS-P3	F	3.59	4.07	7.00	0.74	−0.07	−0.23
SS-P3	S	−0.31	**0.00**	**−0.07**	0.34	−0.38	−0.39
SS-P6	B	−0.34	**−0.05**	**−0.14**	0.31	−0.39	−0.40
SS-P6	F	**0.12**	**0.24**	0.69	0.50	−0.34	−0.24
SS-P7	B	−0.40	**−0.04**	−0.48	**−0.13**	−0.41	−0.43
SS-P7	F	−0.52	−0.39	−0.69	−0.35	−0.41	−0.31
ADAPT	S	−0.77	−0.38	na	na	−0.63	−0.58
TRUE		346	0.117	29	0.102	356	834

NOTE: For the first four management statistics, values within 25% of the truth are shown in boldface type.

TABLE 5.6B Data Set 2: Summary of Relative Error [(estimated − true)/true] in Important Management Parameters—Results with $M = 0.225$

Model	Data source	EB_{30}	$\dfrac{EB_{30}}{EB_1}$	TAC_{31}	$\dfrac{TAC_{31}}{\widehat{EB}_{31}}$	$\bar{R}$	$\bar{S}$
DDKF	F	5.88	3.16	10.00	0.60	−0.48	na
DDKF	S	−0.35	0.29	**−0.10**	0.39	−0.67	na
ADMB4	F	**0.07**	0.66	1.28	1.12	−0.25	−0.32
ADMB4	B	−0.49	**0.02**	**0.00**	0.96	−0.28	−0.39
NRC ADAPT	F	1.92	2.95	6.37	1.53	0.00	0.26
NRC ADAPT	S	**−0.05**	**−0.10**	0.27	0.33	−0.19	0.24
NRC ADAPT	B	0.26	**0.12**	0.62	0.29	−0.18	0.24
ASA	S	0.64	0.84	0.84	**0.12**	−0.21	−0.22
SS-P3	B	**−0.04**	**0.17**	0.50	0.56	−0.24	−0.32
SS-P3	F	4.14	3.63	8.33	0.81	0.13	−0.11
SS-P3	S	−0.27	**−0.13**	**0.04**	0.43	−0.26	−0.32
SS-P6	B	−0.29	**−0.15**	**0.00**	0.41	−0.27	−0.32
SS-P6	F	**0.15**	**0.03**	0.78	0.55	−0.21	−0.13
SS-P7	B	−0.36	**−0.12**	−0.38	**−0.04**	−0.29	−0.37
SS-P7	F	−0.54	−0.52	−0.73	−0.41	−0.29	−0.22
TRUE		346	0.117	29	0.102	356	834

NOTE: For the first four management statistics, values within 25% of the truth are shown in boldface type.

TABLE 5.7A   Data Set 3: Summary of Relative Error [(estimated – true)/true] in Important Management Parameters—Results With Estimated M

Model	Data source	EB_{30}	$\dfrac{EB_{30}}{EB_1}$	TAC_{31}	$\dfrac{TAC_{31}}{\widehat{EB}_{31}}$	$\bar{R}$	$\bar{S}$
DD	B	4.62	2.63	na	na	–0.07	na
DD	F	14.36	5.11	13.73	**–0.04**	na	na
DD	S	0.31	0.53	**0.07**	**–0.18**	na	na
ADMB1	B	0.39	1.34	na	na	0.01	–0.28
ADMB1	F	7.47	5.52	na	na	0.53	0.61
ADMB1	S	0.93	1.80	na	na	–0.01	–0.20
ADMB2	B	0.27	1.84	na	na	0.00	–0.28
ASA	B	1.93	2.79	3.44	0.52	–0.18	–0.05
ASA	F	5.99	5.72	11.76	0.82	0.10	0.38
ASA	S	0.61	1.07	1.14	0.33	–0.28	–0.12
SS-P3	B	2.31	2.82	3.70	0.42	0.13	–0.04
SS-P3	F	4.63	4.05	6.99	0.42	0.39	0.30
SS-P3	S	0.78	1.02	1.33	0.31	–0.06	–0.11
SS-P6	B	0.87	1.14	1.47	0.32	–0.05	–0.11
SS-P6	F	2.29	1.99	3.51	0.37	0.17	0.16
SS-P7	B	0.67	1.07	0.93	**0.15**	–0.10	–0.15
SS-P7	F	1.95	2.07	2.47	**0.18**	0.06	0.09
SS-P3A	B	2.80	3.20	4.56	0.46	0.21	–0.02
SS-P3A	F	4.61	4.03	6.94	0.42	0.38	0.30
SS-P3A	S	0.46	0.58	0.84	0.26	–0.10	–0.11
SS-P6A	B	0.47	0.60	0.87	0.27	–0.10	–0.11
SS-P6A	F	2.32	2.03	3.57	0.38	0.18	0.15
SS-P7A	B	**0.17**	0.38	**0.21**	**0.04**	–0.16	–0.16
SS-P7A	F	2.32	2.45	2.99	**0.20**	0.10	0.12
ADAPT	S	**0.25**	2.16	na	na	–0.51	–0.38
TRUE		903	0.156	70	0.094	853	1,322

NOTE: For the first four management statistics, values within 25% of the truth are shown in boldface type.

NRC ADAPT) had at least two statistics within 25% of the truth. All four statistics met the criterion for the ADMB4(B), SS-P3(S), SS-P3(B), SS-P6(B), and the SS-P7(B) models (Table 5.8B). For data set 5, all models achieved the criterion for at least two statistics, except for the ADMB4(B), NRC ADAPT, SS-P3(B), and SS-P3(F) models (Table 5.9B); SS-P6(F), SS-P6(B), SS-P7(B), and SS-P7(F) fell within 25% for all four statistics.

Overall, Tables 5.5B-5.9B show that errors in exploitable biomass and TAC are less than 25% (bold entries) more frequently for complex models (defined here as ADMB4, SS-P6, and SS-P7) than for simple models (defined here as DDKF, NRC ADAPT, ASA, and SS-P3); bold entries for estimates of exploitable biomass occur for 23% (44%) of simple (complex) model runs; bold entries for estimates of TAC occur for 17% (35%) of simple (complex) model runs. Table 5.10 suggests some modest improvement in management parameters by having the true average M; for example, the percent success (as defined in Table 5.10) was greater for two to four of the five data sets (depending on the variable) when M was set to its true mean. Yet, it should be noted that no one model performed superbly in all cases and for all management parameters.

Effect of M Value

The results from these comparisons suggest that having the correct value of M did not significantly improve the assessment results for the ASA and SS models used before and after the true average M value was revealed. This conclusion appears to be independent of which summary statistic or which data sources are used in the assessment. The reason for this modest effect of incorrect M on all models is probably that variability in M, ageing

TABLE 5.7B   Data Set 3: Summary of Relative Error [(estimated − true)/true] in Important Management Parameters—Results with $M = 0.225$

Model	Data source	EB_{30}	$\dfrac{EB_{30}}{EB_1}$	TAC_{31}	$\dfrac{TAC_{31}}{\widehat{EB}_{31}}$	$\bar{R}$	$\bar{S}$
DDKF	F	11.50	5.49	16.01	0.36	−0.06	na
DDKF	S	**−0.03**	0.52	**−0.13**	**−0.10**	−0.60	na
ADMB4	F	3.44	4.75	9.11	1.28	0.53	0.18
ADMB4	B	0.54	1.68	1.97	0.93	0.12	−0.10
NRC ADAPT	F	2.64	6.50	13.62	3.01	0.81	2.02
NRC ADAPT	S	1.25	1.22	2.75	0.67	0.23	1.22
NRC ADAPT	B	1.53	1.48	3.28	0.69	0.26	1.27
ASA	S	1.05	0.73	1.88	0.40	0.16	0.22
SS-P3	B	2.98	2.81	4.89	0.48	0.46	0.12
SS-P3	F	5.07	3.52	7.99	0.48	0.69	0.46
SS-P3	S	1.06	0.95	1.84	0.38	0.17	0.02
SS-P6	B	1.24	1.14	2.13	0.40	0.20	0.02
SS-P6	F	2.32	1.43	3.68	0.41	0.39	0.32
SS-P7	B	1.03	1.11	1.48	**0.22**	0.13	−0.02
SS-P7	F	1.63	1.26	2.12	**0.19**	0.24	0.21
SS-P3A[a]	B	3.58	3.18	5.95	0.52	0.55	0.15
SS-P3A	F	5.19	3.61	8.17	0.48	0.71	0.46
SS-P3A	S	0.62	0.45	1.15	0.33	0.10	0.02
SS-P6A	B	0.59	0.43	1.11	0.32	0.10	0.01
SS-P6A	F	2.19	1.34	3.49	0.41	0.38	0.31
SS-P7A	B	0.28	0.27	0.41	**0.10**	0.03	−0.04
SS-P7A	F	1.54	1.18	2.00	**0.18**	0.22	0.20
TRUE		903	0.156	70	0.094	853	1,322

NOTE: For the first four management statistics, values within 25% of the truth are shown in boldface type.
[a]Two catchability parameters were used, corresponding to the change of survey vessel.

error, changes in catchability and selectivity, and random variability dominated assessment results more so than the choice of a specific M value. The other reason is that many analyses used an estimated M value not too far from the true average M. One notable exception was the application of the ADAPT model to these simulated data sets. Dramatic underestimation of biomass occurred using an M of 0.15. When the correct average M was used, the NRC ADAPT results were better than the original ADAPT results but still showed major departures from the true values. Therefore, other factors still dominated the assessment results.

The committee expected that relative measures of biomass and exploitation would be estimated more accurately than absolute quantities. Examination of Tables 5.5 to 5.9 shows that this expectation was not fulfilled often. The estimated decline or increase in population over the time period (EB_{30}/EB_1) often had as much or more error than the most recent estimate of exploitable biomass (EB_{30}). Similarly, the projected exploitation fraction in year 31 ($TAC_{31}/\widehat{EB}_{31}$) frequently had greater error than the projected catch limit (TAC_{31}) itself. When this occurred, the reason was often that exploitable biomass was underestimated but TAC was overestimated, which resulted in even greater error in the exploitation fraction.

Additional Analyses

Ability of Models to Detect Abundance Trends

Visual comparison of biomass trends from the assessment models in Appendix I (Figures I.1 to I.10) shows that in many cases the lines (1) do not coincide with the true values and (2) are not parallel with true values. For example, the trends in exploitable and total biomass estimated by ADMB1(F) for data set 2 (Figures I.2 and I.7)

TABLE 5.8A   Data Set 4: Summary of Relative Error [(estimated – true)/true] in Important Management Parameters—Results with Estimated M

Model	Data source	EB_{30}	$\dfrac{EB_{30}}{EB_1}$	TAC_{31}	$\dfrac{TAC_{31}}{\widehat{EB}_{31}}$	$\bar{R}$	$\bar{S}$
SP	F	2.39	na	10.92	2.52	na	na
SP	S	1.70	na	3.92	0.82	na	na
SP	B	1.30	na	9.62	3.61	na	na
SP(iterative reweighting)	B	1.53	na	11.08	3.77	na	na
DD	B	3.56	1.64	na	na	na	na
DD	F	11.45	8.52	8.00	−0.28	na	na
DD	S	1.86	**0.20**	1.15	**−0.25**	−0.20	na
ADMB1	B	**0.14**	**−0.07**	na	na	−0.18	−0.06
ADMB1	F	**−0.01**	**−0.18**	na	na	−0.12	0.20
ADMB1	S	0.84	0.63	na	na	−0.06	0.05
ADMB2	B	0.48	**0.23**	na	na	−0.18	−0.06
ASA	B	0.85	0.64	1.15	**0.16**	−0.11	0.13
ASA	F	1.18	0.66	1.69	**0.24**	−0.10	0.29
ASA	S	0.69	0.48	0.85	**0.09**	−0.11	0.14
SS-P3	B	**0.07**	0.27	**−0.08**	**−0.14**	−0.14	−0.11
SS-P3	F	0.35	0.31	**0.23**	**−0.09**	−0.14	0.04
SS-P3	S	**−0.02**	**0.13**	−0.23	**−0.22**	−0.14	−0.09
SS-P6	B	**−0.06**	**0.08**	−0.31	−0.26	−0.14	−0.09
SS-P6	F	**0.00**	**−0.07**	−0.31	−0.31	−0.15	0.08
SS-P7	B	**−0.01**	**0.18**	−0.31	−0.30	−0.15	−0.11
SS-P7	F	**−0.09**	**−0.08**	−0.46	−0.41	−0.15	0.01
ADAPT	S	−0.88	−0.81	na	na	−0.42	−0.30
TRUE		115	0.024	13	0.132	406	780

NOTE: For the first four management statistics, values within 25% of the truth are shown in boldface type.

TABLE 5.8B   Data Set 4: Summary of Relative Error [(estimated – true)/true] in Important Management Parameters—Results With $M = 0.225$

Model	Data source	EB_{30}	$\dfrac{EB_{30}}{EB_1}$	TAC_{31}	$\dfrac{TAC_{31}}{\widehat{EB}_{31}}$	$\bar{R}$	$\bar{S}$
DDKF	F	6.34	6.63	8.69	0.32	−0.56	na
DDKF	S	1.71	**0.12**	3.15	0.53	−0.41	na
ADMB4	F	**0.20**	**0.21**	0.31	**0.09**	−0.02	0.01
ADMB4	B	**−0.07**	**0.09**	**0.15**	**0.24**	−0.03	−0.08
NRC ADAPT	F	7.89	5.94	11.90	0.45	0.20	1.26
NRC ADAPT	S	0.96	0.73	1.61	0.33	0.06	1.24
NRC ADAPT	B	2.16	1.40	2.17	**0.00**	0.07	1.25
ASA	S	0.77	0.33	1.86	0.62	0.03	0.27
SS-P3	B	**0.15**	**0.18**	**0.11**	**−0.03**	0.00	−0.02
SS-P3	F	0.38	**0.22**	0.38	**0.00**	0.00	0.10
SS-P3	S	**0.00**	**0.01**	−0.13	**−0.13**	−0.01	0.00
SS-P6	B	**−0.06**	**−0.04**	−0.21	**−0.16**	−0.01	−0.01
SS-P6	F	**−0.14**	−0.22	−0.41	−0.32	−0.03	0.09
SS-P7	B	**0.04**	**0.08**	−0.21	**−0.24**	−0.01	−0.02
SS-P7	F	**−0.09**	−0.18	−0.44	−0.38	−0.02	/0.08
TRUE		115	0.024	13	0.132	406	780

NOTE: For the first four management statistics, values within 25% of the truth are shown in boldface type.

TABLE 5.9A   Data Set 5: Summary of Relative Error [(estimated − true)/true] in Important Management Parameters—Results with Estimated M

Model	Data source	EB_{30}	$\dfrac{EB_{30}}{EB_1}$	TAC_{31}	$\dfrac{TAC_{31}}{\widehat{EB}_{31}}$	$\bar{R}$	$\bar{S}$
SP	F	0.32	na	−0.31	0.32	na	na
DD	B	0.78	**−0.14**	na	na	na	na
DD	F	4.45	**−0.15**	3.63	**−0.15**	na	na
DD	S	**−0.11**	**−0.22**	−0.35	−0.27	0.11	na
ADMB1	B	−0.37	**−0.22**	na	na	−0.30	−0.30
ADMB1	F	−0.59	−0.67	na	na	−0.38	−0.27
ADMB1	S	−0.40	−0.40	na	na	−0.37	−0.42
ADMB2	B	−0.30	**−0.15**	na	na	−0.28	−0.26
ASA	B	−0.29	−0.37	−0.32	**−0.04**	−0.35	−0.20
ASA	F	−0.35	−0.48	−0.41	**−0.09**	−0.34	−0.12
ASA	S	**−0.16**	**−0.25**	−0.18	**−0.02**	−0.31	−0.13
SS-P3	B	−0.41	−0.39	−0.54	**−0.21**	−0.37	−0.33
SS-P3	F	−0.41	−0.46	−0.55	**−0.23**	−0.33	−0.25
SS-P3	S	−0.31	−0.27	−0.39	**−0.12**	−0.32	−0.27
SS-P6	B	−0.26	**−0.22**	−0.44	**−0.24**	−0.31	−0.25
SS-P6	F	**0.08**	**0.04**	**−0.09**	**−0.16**	−0.13	0.01
SS-P7	B	−0.27	**−0.22**	−0.40	**−0.19**	−0.31	−0.25
SS-P7	F	**0.05**	**0.03**	**−0.15**	**−0.19**	−0.15	0.00
ADAPT	S	−0.93	−0.90	na	na	−0.69	−0.57
ADAPT	B	−0.87	−0.81	na	na	−0.69	−0.56
TRUE		276	0.088	23	0.109	309	720

NOTE: For the first four management statistics, values within 25% of the truth are shown in boldface type.

TABLE 5.9B   Data Set 5: Summary of Relative Error [(estimated − true)/true] in Important Management Parameters—Results with $M = 0.225$

Model	Data source	EB_{30}	$\dfrac{EB_{30}}{EB_1}$	TAC_{31}	$\dfrac{TAC_{31}}{\widehat{EB}_{31}}$	$\bar{R}$	$\bar{S}$
DDKF	F	−0.32	**−0.20**	**−0.23**	**0.14**	−0.88	na
DDKF	S	−0.36	**−0.17**	−0.34	**0.04**	−0.70	na
ADMB4	F	−0.30	**−0.18**	**−0.06**	0.34	−0.08	−0.17
ADMB4	B	−0.53	−0.36	−0.32	0.45	−0.22	−0.32
NRC ADAPT	F	−0.47	**0.22**	0.30	1.43	0.20	0.24
NRC ADAPT	S	−0.42	1.16	1.09	2.62	0.68	1.03
NRC ADAPT	B	−0.42	1.12	1.06	2.54	0.75	1.15
ASA	S	**0.17**	**−0.11**	**−0.02**	**−0.16**	0.15	0.27
SS-P3	B	−0.30	−0.34	−0.41	**−0.16**	−0.14	−0.18
SS-P3	F	−0.30	−0.42	−0.43	**−0.18**	−0.08	−0.07
SS-P3	S	**−0.17**	**−0.20**	−0.29	**−0.13**	−0.08	−0.11
SS-P6	B	**−0.11**	**−0.14**	**−0.22**	**−0.13**	−0.05	−0.07
SS-P6	F	**0.17**	**0.03**	**0.03**	**−0.12**	0.16	0.20
SS-P7	B	**−0.11**	**−0.13**	**−0.23**	**−0.14**	−0.05	−0.07
SS-P7	F	**0.08**	**−0.01**	**−0.08**	**−0.15**	0.12	0.16
TRUE		276	0.088	23	0.109	309	720

NOTE: For the first four management statistics, values within 25% of the truth are shown in boldface type.

TABLE 5.10   Number of Times That 25% Criterion Was Achieved

Data	Results with Estimated M				Results with $M = 0.225$			
	EB_{30}	$\dfrac{EB_{30}}{EB_1}$	TAC_{31}	$\dfrac{TAC_{31}}{\hat{EB}_{31}}$	EB_{30}	$\dfrac{EB_{30}}{EB_1}$	TAC_{31}	$\dfrac{TAC_{31}}{\hat{EB}_{31}}$
1-F	0	2	0	0	1	2	0	0
1-S	2	1	3	2	3	2	1	1
1-B	4	3	2	2	3	3	2	2
% Success	27% (22)[a]	32% (19)	33% (15)	27% (15)	47% (15)	47% (15)	20% (15)	20% (15)
2-F	2	1	0	1	2	1	0	0
2-S	1	3	1	1	1	2	2	1
2-B	2	4	1	1	1	5	2	1
% Success	25% (20)	44% (18)	14% (14)	21% (14)	27% (15)	53% (15)	27% (15)	13% (15)
3-F	0	0	0	3	0	0	0	2
3-S	1	0	1	1	1	0	0	1
3-B	1	0	1	2	0	0	0	2
% Success	8% (25)	0% (25)	11% (19)	32% (19)	7% (15)	0% (15)	0% (15)	33% (15)
4-F	3	3	1	2	3	4	0	2
4-S	1	2	1	3	1	2	1	1
4-B	4	4	1	2	4	4	4	5
% Success	36% (22)	50% (18)	19% (16)	44% (16)	53% (15)	67% (15)	33% (15)	53% (15)
5-F	2	3	2	5	2	5	4	4
5-S	2	2	1	2	2	3	1	3
5-B	0	5	0	4	2	2	2	3
% Success	20% (20)	53% (19)	23% (13)	85% (13)	40% (15)	67% (15)	47% (15)	67% (15)

[a]Numbers in parentheses indicate number of values included in percentage calculations (determined from parameter values given in Tables 5.5-5.9, excluding cells for which "na" is noted).

predict a much larger biomass than actual and a very different trajectory over time. Noncoincident but parallel trends of the estimated quantities may be acceptable for stock assessment purposes because the estimated trend is unbiased despite the error in estimation of absolute abundance. That is, even if actual stock abundance values are unknown, it is useful to be able to detect relative increases and decreases of stock abundance over time. Nonparallel and noncoincident trends are a special problem because neither the stock abundance nor the way this abundance is changing over time is known.

To evaluate whether the models are useful for detecting trends in stock abundance, estimates of exploitable biomass from the various models were investigated for possible patterns in parallelism and coincidence with the true exploitable biomass for all five data sets. The committee prefers to use exploitable biomass as a comparative tool because it includes the effects of selectivity. Parallelism was tested first because estimated trends that are not parallel with the true trend cannot be coincident with the true exploitable biomass.

The test for parallelism proceeded as follows. The true exploitable biomass was subtracted from the estimated exploitable biomass from the combination of any one assessment method and data source (F, S, or B) for each of the 30 time periods. In a sense, the resultant values can be considered "residuals" of the estimated exploitable biomass. Under the null hypothesis of parallelism, these residuals should exhibit a random distribution about some arbitrary mean value. This mean value should be zero for exactly coincident parallel lines and should not be significantly different from zero when random variation is present. The null hypothesis of parallelism would be rejected if evidence for trends or cycles were found in the residuals. The presence of cycles or trends was tested using the "runs up and down test" (Manly, 1991, pp. 172–173). The test statistic is based on the number of runs of consecutive positive and negative terms when differences are calculated between successive observations in the time series of the residuals. A significantly small number of runs indicates trends in the residuals over time whereas a significantly large number of runs indicates rapid oscillation that can be interpreted as evidence of serial correlation.

Probability levels for the observed test statistics for each method and data set from assuming a normal distribution[*] for the test statistic are presented in Table 5.11. Cutoff significance levels for the test statistics were set at $\alpha = 0.05$ and $\alpha = 0.01$. Data sets 1, 3, and 4 had the greatest number of cases for which the null hypothesis was rejected. Only one model or data combination (DD2 [F]) had problems with data set 5 compared with 11 combinations for data set 4. The only difference between these two data sets was that the population was increasing in data set 5 and decreasing in data set 4. Inspection of the differences showed that most of the models had difficulty detecting the declining trend in data set 4 at the beginning of the series, with differences between estimated and true exploitable biomass decreasing over time. For data set 5, the differences were more or less random at the beginning of the series with a tendency in some cases for increasing differences toward the end of the series. However, these increases in the latter part of the series were not as extreme (except for DD2 [F]) as those observed at the beginning of the series for data set 4.

Data set 2 appears to have been the least problematic data set for trends in terms of the differences between estimated and true exploitable biomass despite the addition of 30% underreporting to the conditions for data set 1.

There does not appear to be a consistent pattern of which model did better or worse for any particular data set. Only models ADMB4 (F, S, B), SS-P3 (S, B), and SS-P6 (B) did not exhibit any significant trend over all data sets. The ADMB4 model was the most heavily parameterized (about 400 parameters) and allowed for variable selectivity and catchability. Conversely, both SS-P3 and SS-P6 set selectivity to be constant, and the latter method modeled catchability as a power function of biomass. Despite these differences, all of these models performed well with respect to parallelism whether or not selectivity, catchability, or both varied in the actual data sets.

For those combinations of model and data set for which the null hypothesis was not rejected in Table 5.11, the null hypothesis that mean difference between estimated and true exploitable biomass was not significantly different from zero was tested using a Student's t-test (Table 5.12). In all but one case (data set 4, F), the mean differences for model ADMB4 were significantly different from zero. Therefore, although the results in Table 5.11 indicate that this model generally performed better than the other models in capturing the trend in exploitable biomass, the estimated trend was almost always an underestimate of the true trend except for data set 3 (F), for which the exploitable biomass was overestimated.

Overall, the mean differences were negative for most models in Table 5.12 when the differences were significantly different from zero and no model stands out as having smaller absolute mean differences than the others.

Effects of Ageing Error

Analysts were not asked to undertake analyses of the effect of ageing error on the results, because of time limitations. Instead, the committee constructed a simple age-structured analysis similar to the separable ASA (ASA and SS-P). The following were used: data set 1 only, no survey information, true average value of M, a logistic function for fishery selectivity, and a plus group starting at age 10. A lognormal objective function was used for catch-age and for fishing mortality deviations from fishing effort under constant catchability along the lines of Deriso et al. (1985). In the first analysis, ageing error was ignored and a standard analysis was done. In the second analysis, the model's catch-age was transformed to a perceived catch-age with ageing error included by

---

[*]This was assumed to be valid because a randomization version of the test produced similar results to those using an assumption of normality.

TABLE 5.11   P-Levels for Runs-up Test for Estimated Minus Actual Exploited Biomass

Model	Data Set				
	1	2	3	4	5
ADMB1(F)	*	0.152	*	0.152	0.086
ADMB1(S)	*	0.086	0.178	*	0.105
ADMB1(B)	*	0.105	0.236	0.086	0.178
ADMB2(B)	**	0.086	0.086	0.086	0.054
ADMB4(F)	0.086	0.086	0.086	0.105	0.178
ADMB4(S)	0.086	0.086	0.236	0.236	0.178
ADMB4(B)	0.086	0.086	0.236	0.236	0.178
ADAPT(S)	*	0.235	**	0.085	0.178
DD2(F)	**	*	*	*	**
DD2(S)	0.086	0.236	*	**	0.086
DD2(B)	0.236	0.236	*	**	0.236
DD3(F)	0.236	*	*	0.152	0.152
DD3(S)	0.236	0.086	0.086	**	0.178
ASA (F)	**	*	*	**	0.236
ASA (S)	*	0.236	0.178	**	0.236
ASA (B)	**	0.236	0.236	**	0.236
SS-P3(F)	**	0.084	0.236	*	0.236
SS-P3(S)	0.236	0.086	0.236	0.152	0.236
SS-P3(B)	0.152	0.086	0.236	0.152	0.236
SS-P6(F)	*	0.086	0.236	0.152	0.236
SS-P6(B)	0.236	0.086	0.236	0.152	0.236
SS-P7(F)	0.236	0.086	0.105	*	0.236
SS-P7(B)	0.236	0.236	0.178	*	0.236

NOTE: * indicates that test statistic was significant at the $\alpha = 0.05$ level and ** indicates significance at the $\alpha = 0.01$ level. All significant test statistics were less than zero, indicating evidence for trend.

multiplying the matrix of catch-at-age by the ageing error matrix (older ages in the original ageing error matrix were pooled and some exponential weighting of the older ages was done).

Estimates of exploitable biomass from the two analyses are shown in Figure 5.14. There are only minor differences in biomass between the two analyses. The estimates differ substantially from the true values due to the increasing catchability. Finally, the estimates are comparable to other F analyses of data set 1, which suggests that ageing error played a minor role in causing the apparent biases in the analyses.

The committee derived a measure of fishery effort from the survey information by dividing total yield by survey CPUE. When its model was run with this alternate effort measure, there was minimal bias in estimated exploitable biomass, even without correcting for ageing error. This suggests that having a good survey index along with a good estimate of natural mortality can mitigate the bias observed in some analyses of data set 1.

Retrospective Analyses

As discussed in Chapter 3, retrospective analysis is an essential tool for studying the consistency of stock assessment results and methods over time. Retrospective analyses were performed for the three age-structured methods by a subset of the analysts using the Stock Synthesis model, AD Model Builder, and NRC ADAPT. Sixteen independent assessments were required for each data set and data source combination. The 16 independent assessments for a specific combination correspond to assessments utilizing the first 1-15 years of data, the first 1-16 years of data, and so forth, through the complete 1-30 years of data, yielding retrospective data points for years 15-30 for each combination. Completion of the total number of assessments required 5 data sets × 16 year sets × 3 data source combinations = 240 separate analyses for each method. As a practical matter, the analyses were generally completed using automated algorithms that maximized some type of likelihood function. Such

TABLE 5.12   Mean Difference Between Estimated and Actual Exploited Biomass Where Null Hypothesis Was Not Rejected in Table 5.11

Model	\multicolumn{5}{c}{Data Set}
	1	2	3	4	5
ADMB1(F)		1299.7 *	2637.9 **	472.3 **	−659.0 **
ADMB1(S)		−541.3 **	347.3 *		−921.5 **
ADMB1(B)		−595.0 **	249.6	65.3	−986.9 **
ADMB2(B)		−700.0 **	71.5	84.5	−905.8 **
ADMB4(F)	−266.3 **	−532.8 **	630.8 **	41.5	−914.5 **
ADMB4(S)	−418.7 **	−791.0 **	−535.5 **	−243.6 **	−2035.9 **
ADMB4(B)	−415.6 **	−768.2 **	−512.1 **	−233.8 **	−1913.0 **
ADAPT(S)		−961.1 **		−510.6 **	−2581.1 **
DD2(F)					
DD2(S)	663.4 **	−266.8 **			259.8
DD2(B)	921.9 **	−163.2 *			3782.7 **
DD3(F)	762.1 **			1194.6 **	−1572.2 **
DD3(S)	322.9 **	−635.7 **	107.4		−715.5 **
ASA (F)					180.3
ASA (S)		147.2 **	−9.5		103.1
ASA (B)		192.9 **	293.6 *		−223.2
SS-P3(F)		−145.0 *	1467.4 **		−690.3 **
SS-P3(S)	−9.1	−504.8 **	−40.3 **	−14.7	−915.0 **
SS-P3(B)	−2.5	−504.0 **	411.8 **	−67.9	−1139.9 **
SS-P6(F)		−2.5 **	−504.8 **	1467.4 **	411.8 **
SS-P6(B)	−9.1	−145.0 **	−504.0	−40.3	195.7 **
SS-P7(F)	116.6 **	−383.8 **	820.4 **		416.3 **
SS-P7(B)	−38.2	−576.6 **	−203.7 **		−828.8 **

NOTE: * indicates that the mean was significantly different from zero at the $\alpha = 0.05$ level and ** indicates significance at the $\alpha = 0.01$ level (Student's t-test).

automation prevents the kind of detailed inspection normally afforded individual assessments and thus may contribute to some of the retrospective errors discussed below. On the other hand, the analysts were provided the true average value of natural mortality rate used to generate the simulated data sets.

To illustrate the retrospective results, examples of "good" and "bad" retrospective patterns are given. The good pattern is shown in Figure 5.15 for model ADMB4(F) and data set 4. Values of exploitable biomass and estimated coefficients of variation are given for each retrospective run; the circle at the end of each line marks the final year of each retrospective assessment, used to identify which run is plotted. The coefficients of variation are estimated by ADMB from the covariance matrix of the model parameters. The true values for exploitable biomass are also shown. The estimated exploitable biomass is generally near the true value except for the first five runs, in which the assessments consistently overestimate stock biomass in the final years; later data bring the discrepant values close to the truth. The coefficients of variation for the terminal year are relatively similar, and as new data are added, the errors decrease, showing that historical estimates of biomass tend to converge.

The bad pattern is shown in Figure 5.16 for model ADMB4(F) and data set 3. Estimated exploitable biomass is rarely near the true value and there are stretches of many years in which biomass is either over- or under-estimated grossly. Historical estimates of biomass converge much more slowly than in the previous case and do not converge to the true values. The coefficients of variation for the terminal year are fairly similar, as before, but they are much larger.

Another way to examine the results of retrospective analyses is to plot estimates of the final year of successive retrospective assessments with the true values (Figures 5.17 to 5.21). Each line in the figures shows, for a given method, the series of annual estimates of biomass (i.e., values marked by circles in Figures 5.20 and 5.21), that

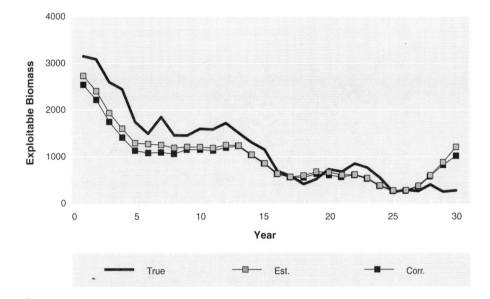

FIGURE 5.14 Effect of ageing error on estimated exploitable biomass (thousand metric tons) for data set 1 using age-structured assessment with the fishery index. NOTE: "Est." denotes values for which ageing error is ignored. "Corr" denotes values in which a correlation for ageing error was made. True values are also shown.

presumably would be used for setting catch limits. Biomass estimates tend to be either above or below the true values for a number of consecutive assessments, indicating persistent bias or autocorrelation in the errors. The average relative and absolute relative deviations between estimated and true terminal year exploitable biomass and the serial correlation between successive deviations can be used to characterize trends in retrospective errors. Table 5.13 shows the average relative deviation by data set and assessment method. These results show that average errors were generally smaller (in absolute terms) for assessments that utilized survey data. In data sets 1-4, which correspond to depletion trajectories, most assessments tended to overestimate biomass; errors were especially large in the assessments that used only fishery CPUE. In addition, errors made in successive assessments tended to be positively correlated (Table 5.14). There was no generally best assessment method in terms of low bias, although the NRC ADAPT method was clearly inferior to others in the case of assessments using fisheries data alone. The models that allow for trends in selectivity and/or catchability tended to perform similarly in terms of the point estimates of terminal year biomass; the exception was data set 5, the recovering stock, in which Stock Synthesis models had a much better performance (Figure 5.21 and Tables 5.13 and 5.14).

Absolute relative deviation is a measure of the closeness of an estimate to its true value. Table 5.15 gives the average absolute relative deviations between estimated and true terminal year exploitable biomass by data set and assessment method or type. When using a criterion of overall average error of 25% or less of the true values as the goal for stock assessment, only 8 of the 50 entries in Table 5.15 (shown in bold) satisfy that criterion. Assessments that utilize survey data perform roughly twice as well as those that do not, at least as far as complex assessment methods (i.e., Stock Synthesis and ADMB) are concerned. The simpler NRC ADAPT method performs nearly as well as complex models overall when survey data are available, but fails, sometimes strikingly, when only fisheries data are available. If instead of average error, the errors made in each of the 16 retrospective assessments (terminal years 15-30) are considered, many assessments failed to achieve the "less than 25% criterion" (Table 5.16), and errors in excess of 50% were frequent. This was the case even for assessments that used survey indices of abundance.

A disturbing feature of all the assessment methods is the tendency to lag in their detection of trends in population abundance over time. Table 5.17 presents correlation coefficients between the relative deviation (as

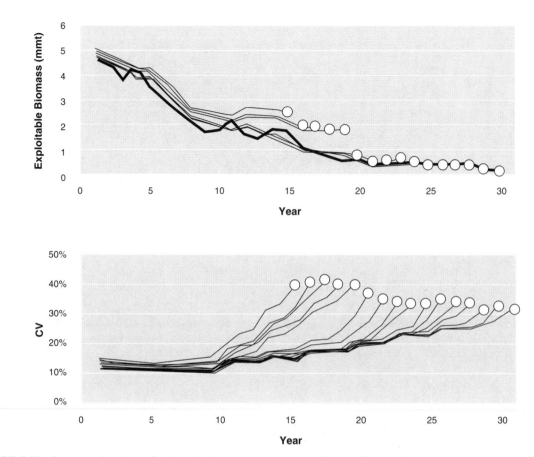

FIGURE 5.15   Retrospective plots of exploitable biomass and corresponding coefficient of variation for model ADMB4(F) and data set 4 (thick line in the top panel denotes true values of exploitable biomass).

calculated for Table 5.14) and the recent change in annual exploitable biomass (as calculated by the difference between exploitable biomass in the terminal year of each run and the previous year). The correlation coefficients are all negative, indicating a tendency for all methods and all types of data to overestimate biomass during periods of population decline and/or to underestimate biomass during periods of population increase. An example of the retrospective error pattern is shown in Figure 5.22. The disproportionately high number of points in quadrants 2 and 4 indicate the tendency for the SS-P3 model (using only fishery data) to overestimate exploitable biomass during periods of decreasing abundance and underestimate exploitable biomass during periods of increasing abundance. Tables 5.18 and 5.19 are included to provide a closer examination of the pattern of errors. These tables show substantial variation among methods and types of auxiliary data in their tendency to underestimate biomass during periods of population increase or decrease. They show that some method or types tend to have one-sided errors (e.g., consistent overestimation on data set 1 by NRC ADAPT with fisheries data only).

IMPLICATIONS OF MODEL RESULTS

The purpose of the simulation study was to probe the performance of commonly used stock assessment models under severe conditions where these models were suspected not to perform well. The data and models tested the effects of ageing error, variation in natural mortality, changes in fishery selectivity and catchability, a lack of proportionality in the relationship between fishery CPUE and biomass, a change in survey selectivity, a dome-shaped selectivity curve for the survey, underreporting of catch, and random variability in the dynamics and

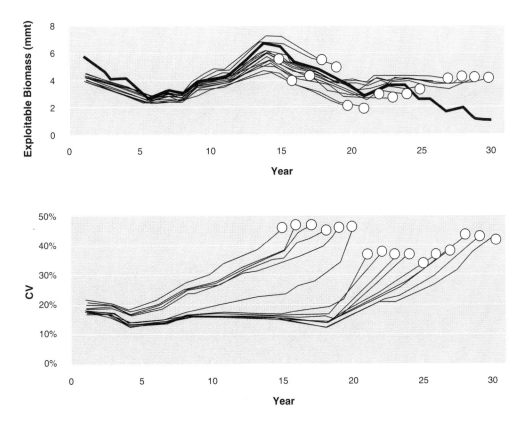

FIGURE 5.16 Retrospective plots of exploitable biomass and corresponding coefficient of variation (CV) for model ADMB4(F) and data set 3 (thick line in the top panel denotes true values of exploitable biomass).

sampling processes. These conditions are known to exist in actual stock assessments, and simulation results confirmed that these co-occurring complications can lead to substantial bias and variability in estimates of population and management parameters.

Because it is rare to know enough about any specific fish population to estimate parameters for all of the processes that can affect that population, its fisheries, and surveys, it is impossible to know if the five simulated populations could be considered typical of any specific actual population. The committee believes, however, that the parameters chosen are plausible and could reflect what actually occurs in some populations. Consequently, actual stock assessments could be substantially inaccurate, suggesting that uncertainty of estimates must be a primary concern in the interpretation of stock assessment results and the derivation of harvesting policies. Only recently have approaches that explicitly treat uncertainty in stock assessment models been developed, as described in Chapters 3 and 4. It is imperative that future assessments incorporate and present realistic measures of uncertainty.

The committee did not construct the simulation study to search for a "Holy Grail" of stock assessment models. Time and personnel constraints limited the simulations to a single replication of five different populations. Complete evaluation of models would require hundreds of simulations of several populations. Such evaluation rarely appears in the published literature because the task is quite daunting and not as interesting to scientists as developing new methods.

Classical forms of production and delay-difference models include a spawner-recruit or stock-production relationship, which allows for direct calculation of productivity parameters such as MSY. One might wonder whether this requirement makes these models more ambitious in what they are trying to achieve and hence puts them at a disadvantage compared with age-structured models. In actuality, it is straightforward to imbed a

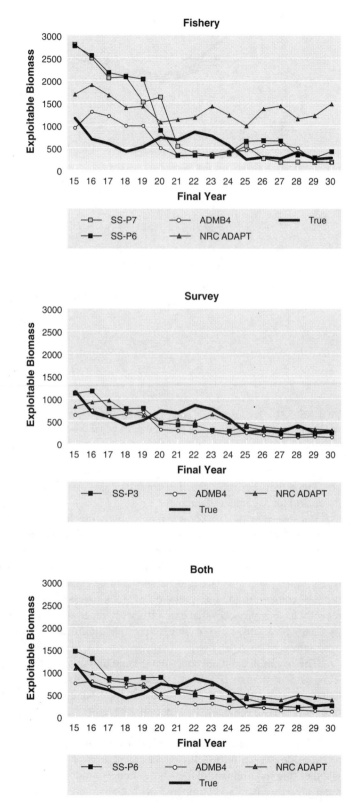

FIGURE 5.17 Comparison of estimated and true biomass (thousand metric tons) in terminal years for 16 retrospective assessments (terminal years 15-30 assessments) conducted on data set 1 using Stock Synthesis (SS-P3, SS-P6, and SS-P7), AD Model Builder, and NRC ADAPT; *M* was fixed at the true average value in all cases.

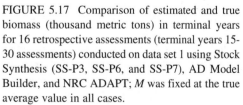

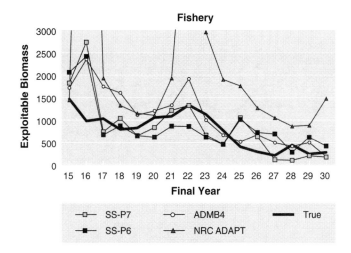

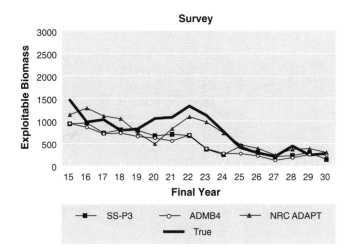

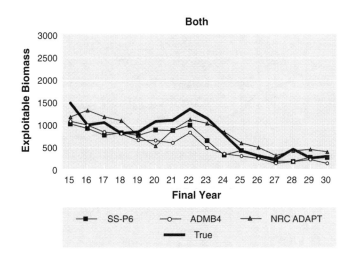

FIGURE 5.18 Comparison of estimated and true biomass (thousand metric tons) in terminal years for 16 retrospective assessments (terminal years 15-30 assessments) conducted on data set 2 using Stock Synthesis (SS-P3, SS-P6, and SS-P7), AD Model Builder, and NRC ADAPT; M was fixed at the true average value in all cases.

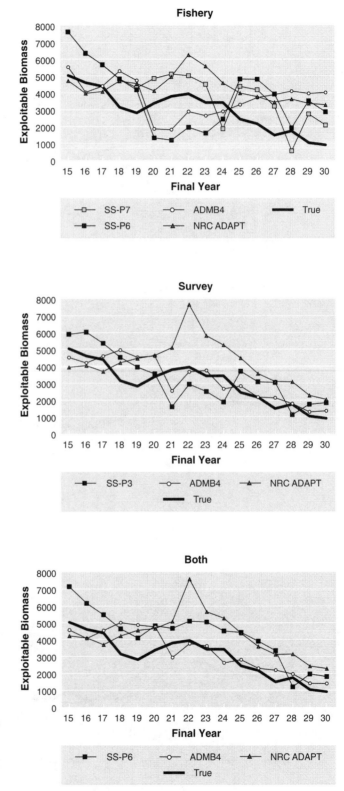

FIGURE 5.19 Comparison of estimated and true biomass (thousand metric tons) in terminal years for 16 retrospective assessments (terminal years 15-30 assessments) conducted on data set 3 using Stock Synthesis (SS-P3, SS-P6, and SS-P7), AD Model Builder, and NRC ADAPT; *M* was fixed at the true average value in all cases.

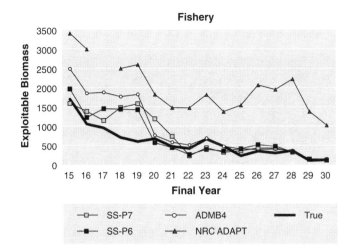

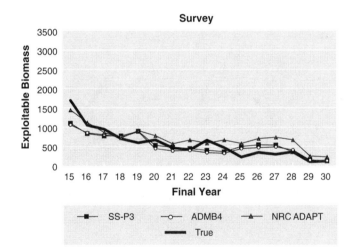

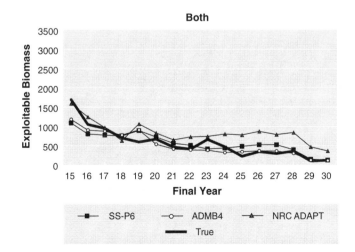

FIGURE 5.20 Comparison of estimated and true biomass (thousand metric tons) in terminal years for 16 retrospective assessments (terminal years 15-30 assessments) conducted on data set 4 using Stock Synthesis (SS-P3, SS-P6, and SS-P7), AD Model Builder, and NRC ADAPT; *M* was fixed at the true average value in all cases.

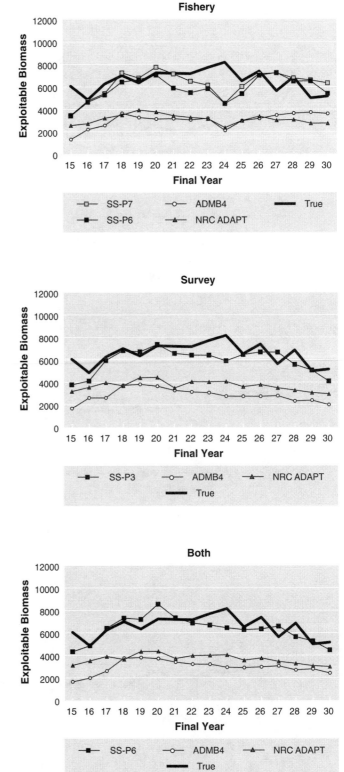

FIGURE 5.21   Comparison of estimated and true biomass (thousand metric tons) in terminal years for 16 retrospective assessments (terminal years 15-30 assessments) conducted on data set 5 using Stock Synthesis (SS-P3, SS-P6, and SS-P7), AD Model Builder, and NRC ADAPT; *M* was fixed at the true average value in all cases.

TABLE 5.13   Average of Relative Deviations Between Estimate and True Exploitable Biomass in Terminal Year (for terminal years 15-30)

Data Set	SS-P6			SS-P7	ADMB4			NRC ADAPT		
	F	B	S	F	F	B	S	F	B	S
1	106	**13**	**−3**	73	27	−28	−31	207	28	**13**
2	40	**−23**	**−23**	**11**	52	−33	−35	251	**13**	**2**
3	51	51	**21**	52	63	**18**	**14**	72	60	57
4	**25**	**19**	**15**	31	52	**−1**	**8**	413	102	58
5	**−9**	**−3**	**−9**	**−3**	−53	−55	−56	−51	−43	−43
Average	43	11	0	33	28	−20	−20	178	32	17

NOTE: Boldface values indicate average deviations not exceeding 25%.

TABLE 5.14   Serial Correlation Between Errors in the Estimates of Exploitable Biomass[a] in the Terminal Year

Data Set	SS-P6			SS-P7	ADMB4			NRC ADAPT		
	F	B	S	F	F	B	S	F	B	S
1	0.86	0.71	0.56	0.89	0.64	0.54	0.47	0.46	0.38	0.41
2	0.26	0.21	0.46	0.02	0.32	0.51	0.56	—	0.32	0.27
3	0.66	0.04	0.42	−0.12	0.72	0.42	0.30	0.71	0.66	0.68
4	0.56	0.54	0.39	0.64	0.75	0.22	0.32	—	0.41	0.40
5	0.39	0.10	−0.04	0.38	0.37	0.07	0.00	0.30	0.08	0.07

[a]Deviations between estimated and true exploitable biomass (for terminal years 15-30).

TABLE 5.15   Average of Relative Absolute Deviations Between Estimate and True Exploitable Biomass in Terminal Year (for terminal years 15-30)

Data Set	SS-P6			SS-P7	ADMB			NRC ADAPT		
	F	B	S	F	F	B	S	F	B	S
1	133	40	40	114	65	44	44	207	38	32
2	75	**24**	28	54	56	33	36	251	30	**23**
3	83	55	45	66	84	26	**24**	76	65	63
4	44	34	35	48	52	**22**	34	388	104	63
5	**18**	**11**	**13**	**15**	53	55	56	51	43	43
Average	71	33	32	60	62	36	40	195	56	45

NOTE: Relative deviation = (estimated − true)/true; Relative absolute deviation = |relative deviation|.

TABLE 5.16A Number of Assessments with Estimates of Terminal Exploitable Biomass Within ± 25% of True Value

Data Set	SS-P6 B	SS-P6 F	SS-P6 S	ADMB4 B	ADMB4 F	ADMB4 S	NRC ADAPT B	NRC ADAPT S	NRC ADAPT F
1	6	2	5	3	4	3	4	7	0
2	9	6	8	6	7	5	8	11	2
3	2	0	3	10	5	10	3	3	4
4	9	5	9	11	7	7	6	7	0
5	16	12	14	0	1	0	0	0	0

NOTE: Maximum $n = 16$.

TABLE 5.16B Number of Assessments with Relative Errors in Estimates of Terminal Exploitable Biomass Within ± 50% of True Value

Data Set	SS-P6 B	SS-P6 F	SS-P6 S	ADMB4 B	ADMB4 F	ADMB4 S	NRC ADAPT B	NRC ADAPT S	NRC ADAPT F
1	12	6	8	9	6	8	12	13	2
2	14	4	14	13	10	13	14	15	3
3	11	7	11	13	8	14	6	6	7
4	13	10	6	14	11	14	7	8	0
5	16	16	16	5	6	5	15	14	8

TABLE 5.17 Pearson Correlation Between Relative Deviation and True Change in Exploitable Biomass for Terminal Years 16-30

Data Set	SS-P3 F	SS-P3 B	SS-P3 S	SS-P7 F	ADMB4 F	ADMB4 B	NRC ADAPT F	NRC ADAPT B	NRC ADAPT S
1	−0.37	−0.48	−0.53	−0.39	−0.34	−0.36	**−0.17**	−0.55	−0.59
2	−0.48	**−0.31**	−0.33	−0.60	−0.35	−0.42	−0.52	−0.57	−0.64
3	−0.53	−0.51	−0.58	−0.46	−0.29	−0.37	−0.17	**−0.11**	−0.13
4	−0.28	−0.05	−0.10	−0.27	−0.40	−0.26	−0.03	**−0.02**	0.02
5	−0.51	**−0.38**	−0.49	−0.47	−0.46	−0.45	**−0.38**	−0.48	−0.46
Average	−0.43	−0.35	−0.41	−0.44	−0.36	−0.37	−0.25	−0.35	−0.36

NOTE: Change in biomass is between terminal year and previous year; the lowest coefficient for each data set is noted in boldface type.

TABLE 5.18 Percent of Positive Relative Deviations Given a Positive True Change (labeled as an "up") in Exploitable Biomass for Terminal Years 16-30 Assessments

Set	# Ups	SS-P3 F	SS-P3 B	SS-P3 S	SS-P7 F	ADMB F	ADMB B	NRC ADAPT F	NRC ADAPT B	NRC ADAPT S
1	6	67%	50%	17%	33%	50%	17%	100%	67%	50%
2	7	14%	0%	0%	14%	86%	0%	100%	29%	14%
3	5	20%	80%	20%	60%	20%	40%	100%	100%	100%
4	5	20%	80%	60%	40%	80%	20%	100%	100%	80%
5	8	13%	38%	13%	38%	0%	0%	0%	0%	0%
Average	6.2	27%	50%	22%	37%	47%	15%	80%	59%	49%

NOTE: Change in biomass is between terminal year and previous year. Percent positive deviations below 50% indicate a tendency of the method to underestimate biomass during periods of population increase. Relative deviation = (estimate−true)/true exploitable biomass for terminal year.

TABLE 5.19 Percent of Negative Relative Deviations Given a Negative True Change (labeled as a "down") in Exploitable Biomass for Terminal Years 16-30 Assessments

Set	#Downs	SS-P3 F	SS-P3 B	SS-P3 S	SS-P7 F	ADMB F	ADMB B	NRC ADAPT F	NRC ADAPT B	NRC ADAPT S
1	9	33%	56%	56%	56%	44%	67%	0%	33%	33%
2	8	25%	63%	50%	50%	25%	100%	0%	13%	25%
3	10	10%	0%	10%	0%	20%	10%	20%	20%	20%
4	10	30%	30%	30%	20%	10%	50%	0%	10%	10%
5	7	57%	29%	57%	57%	100%	100%	100%	100%	100%
Average	8.8	31%	35%	41%	37%	40%	65%	24%	35%	38%

NOTE: Change in biomass is between terminal year and previous year. Percent negative deviations below 50% indicate a tendency of the method to over-estimate biomass during periods of population decrease. Relative deviation = (estimate−true)/true exploitable biomass for terminal year.

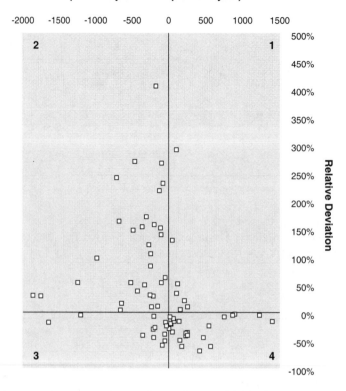

FIGURE 5.22 Example of retrospective error pattern (estimates from model SS-P3[F]) over assessments for terminal years 16-30 and all data sets.

spawner-recruit relationship into age-structured models, as shown in Deriso et al. (1985), which could then be used to calculate MSY and other such parameters. In ADMB, a recruitment model is specified which can include a spawner-recruit relationship. Age-structured models are able to reconstruct annual recruitment values without including a spawner-recruit relationship in the assessment model. In this study, analysts using the delay-difference model did not use a spawner-recruit relationship but instead used a recruitment index derived from the survey. Although the production model used a logistic stock-production relationship, one can construct a form of this model which incorporates a recruitment index. (This was mentioned as a possibility by the analyst but not used, because a recruitment index requires age-structured information and the goal was to see what could be extracted from a simple model). For management, some biological reference points (BRPs, such as $F_{40\%}$) related to spawning biomass per recruit can be calculated without using a spawner-recruit relationship, and such BRPs are being used with increasing frequency in actual assessments. But the spawner-recruit relationship should be investigated to evaluate whether the BRP is appropriate. Whether it is better to do so by incorporating the relationship within the assessment model or as a separate analysis using output from the assessment model is an open question.

Because a full evaluation of stock assessment models could not be done, the committee used its resources to explore the performance of the models in a simulated stock assessment setting in an attempt to (1) evaluate existing methods and (2) suggest new directions of research into stock assessment methods. It is obvious that a more comprehensive evaluation of stock assessment methods should be undertaken, given the results of this study. Issues related to the treatment of measurement and process errors, the functional dependence of population parameters (e.g., catchability) on biomass, the choice and weighting of individual data sets, information conflicts among data sets, and the appropriate level of model complexity are all unresolved at present, and require attention if there is to be greater confidence in stock assessments. These would be fruitful areas of research.

The committee's analysis indicates that high-quality data, fundamentally the availability of reliable indices to calibrate the models, are essential to produce reliable abundance estimates. In most cases, use of the fishery abundance index resulted in poor performance unless the model contained additional parameters to deal with the trend in the index.

Surplus production and delay-difference models did not perform as well overall as age-structured models; this is not surprising because the simulated data were designed for use with age-structured methods. Surplus production models require a straightforward and immediate response of the population to changes in harvesting levels. The simulated populations were more affected by recruitment fluctuations than by changes in harvest levels. The corruption of indices of abundance by catchability and selectivity changes and by underreporting of catch would make stock assessment with surplus production models nearly impossible. Better results were achieved for delay-difference models because analysts utilized an index of recruitment from the survey and/or fishery data, rather than relying on a stock-recruitment model. Using a knife-edge selectivity assumption in these models when there was an underlying selectivity pattern with age increased the uncertainty and potential bias in estimates of population parameters. Nevertheless, better results were obtained for these models when the survey index was used alone than when only the fishery index or both indices were used.

Among the age-structured models, simple models such as ASA, SS-P3, or NRC ADAPT performed reasonably well when only the survey index was used and when the dynamics of the population and harvest were not too complex. More complex models such as SS-P6, SS-P7, and ADMB4 were sometimes able to handle more complex dynamics and indices with trends. However, the success of these more complicated models depended on correct specification of the dynamic changes in selectivity, catchability, and natural mortality.

Simulation results suggest that models with greater complexity offer promise for improving stock assessment. The Kalman filter (in DDKF) and generalized parametric approach (in AD Model Builder) allowed more realistic treatment of process and measurement errors. The Bayesian treatment of parameters (also in AD Model Builder) provided a means for incorporating uncertainty directly into the analysis and yielded results in terms of posterior probability distributions, which explicitly presented uncertainty. The incorporation of functional dependence of catchability and flexibility in model specification (in SS-P6 and SS-P7) provided a more deterministic way of adding realism. Although no specific model outperformed others in the simulations, the committee was intrigued with how more complex models could reduce, at least partially, the biasing effects related to fishery catchability and selectivity changes.

Simulation results showed that when there is substantial recruitment variability, production models do not perform well. Only with populations that exhibit a strong negative response to fishing should these models be used for routine assessment. Nevertheless, there will be situations in which data limitations preclude the use of other methods. Delay-difference models fared better than production models but worse than age-structured models. Although delay-difference models might be used in situations in which ageing is subject to great error or not possible, it would be more prudent to utilize the age or length information in stock assessments. One of the reasons delay-difference models performed as well as they did in the simulations was use of a recruitment index from the survey. Thus, the development of recruitment indices for use in stock assessments should be considered.

Although the construction of better stock assessment models is likely to lead to better assessments, accurate and precise information about the population is of paramount importance. Although this conclusion is fairly obvious ("garbage in-garbage out"), the simulation study provided a clear illustration of the importance of good data and information. In data set 4, there were few violations of the underlying assumptions used in the assessments, and not surprisingly, most of the stock assessment models performed acceptably. The worst results for this data set were obtained when only the fishery index was used, showing that bad data weaken a stock assessment. Each of the other data sets had some additional complicating factor, that resulted in poorer results than for data set 4. Hence, poor information (e.g., not knowing about a change in catchability) is an additional factor that weakens a stock assessment. When the combination of poor data and poor information becomes large enough (e.g., data set 3), it can be almost impossible to extract any useful stock assessment information.

The major conclusion from the simulation study was that a good index of abundance is needed for useful stock assessment information, not that fishery indices should not be used. Much effort is required to validate any index as a measure of abundance.

Specific examples of poor information incorporated into the simulation study deserve further comment. The misspecification of natural mortality has long been recognized as a critical problem in stock assessment. Overestimation of natural mortality leads to overestimation of both population abundance and optimal harvesting rate due to yield-per-recruit consequences. In the simulation study, natural mortality was rarely overestimated. Rather, the variability in natural mortality made it difficult to estimate other parameters. Two possible approaches to incorporating variable natural mortality are (1) to pursue methods such as multispecies virtual population analysis (MSVPA), that utilize data on food habits of different species, and (2) to use Bayesian specification of priors for natural mortality, provided that appropriate priors can be found (e.g., through meta analysis).

Underreporting in data set 2 led to underestimation of population biomass, with greater effects at the beginning of the data series. Thus, the proportionate decline in population over time was underestimated, which could lead managers to think that less strict harvesting policies are adequate to rebuild a depleted population. The obvious solution is to design the catch-reporting system so that underreporting is less likely to occur (see Chapter 2). It is notable that underreporting can work to the detriment of fishers in the long run by corrupting the data used in assessment models.

The decline in age selectivity in data sets 1, 2, and 3 resulted in a fishery index that would be higher than under constant selectivity, because more younger fish would be harvested and added to the fishery index. Such a change could occur by targeting of smaller fish as a population became depleted or by a change in the age selectivity curve due to density-dependent or environmentally induced changes in growth. Collection and analysis of growth data are critical to understanding changes in age selectivity. If changes can be detected from such observations, modeling can readily account for them, either by utilizing length-based methods or by having separate sets of age selectivity parameters (e.g. in SS-P7 or ADMB).

The change in survey catchability in data set 3 created a situation in which neither index of abundance was proportional to biomass. When a change in catchability was incorporated into the models, its value could be estimated in many cases, with a resultant improvement in stock assessment results. If the committee had not told the analysts of the potential change, however, it is doubtful whether it would have been detected. The implication is that calibration of survey catchability is an important consideration; calibration studies should be done when there are changes in vessels, crews, or operations that affect the way a survey is conducted (see ASMFC, 1997).

Retrospective analyses from the simulation revealed that stock assessments can vary substantially from the true values over time. These analyses also illustrated that the departures of estimated stock biomass from true values can persist in one direction over time. Consequently, management actions could have deleterious effects on the population long before they are observed. Thus, conservative harvest policies should be developed. The committee encourages greater use of retrospective analysis in stock assessment. The results can suggest when model misspecification is occurring and when data sources are providing contradictory information. The retrospective analysis reported herein focused on an evaluation of modeling. Another type of retrospective analysis involves summarization of previous stock assessment results regardless of which model was used. This type of retrospective analysis can be useful in examining the stability of the stock assessment process as actually implemented.

The additional investments in stock assessment research recommended by the committee require additional commitments of personnel, field research, and analytical research. Chapter 4 contains a discussion of where investments should be made—not merely in collecting more of the same information but in improving the type of information collected as well. Simulation results suggest that if such investments cannot be made, some stock assessments will be far from the truth periodically, and consequently that management mistakes leading to fish population collapses and other negative consequences will be made.

The approach of having a committee conduct simulation research involving NMFS scientists in an independent review of stock assessment methods is not undertaken frequently. This approach was quite useful for brainstorming issues related to stock assessment, and new understanding and novel approaches to stock assessment were inspired by this project. This work was designed to evaluate not the individual NMFS scientists involved but the methods used in stock assessment. Nevertheless, it took a great deal of courage for these scientists to participate, and the committee was impressed with the analysts' willingness and ability to provide what it requested, their creativity in applying existing methods and developing new ones, and their fortitude in taking on a difficult assignment in addition to their regular duties. To foster excellence in stock assessment, NMFS should continue to support and encourage scientists to engage in creative stock assessment activities (e.g., workshops, gaming sessions, and conferences) so that the process of doing stock assessment does not become routine and stale.

6

Findings and Recommendations

The committee developed the following findings and recommendations based on the information presented in Chapters 1 through 4 and the results of the simulation studies described in Chapter 5. The committee makes specific recommendations about how stock assessments are conducted; data collection and assessment methods; harvest strategies; a rigorous evaluation system; continued development of new assessment techniques; periodic peer review of assessments and assessment methods; and education and training of stock assessment scientists.

HOW SHOULD STOCK ASSESSMENTS BE CONDUCTED AND BY WHOM?

Finding: Stock assessments do not always provide enough information to evaluate data quality and estimate model parameters.

The committee concluded that stock assessments are sometimes incomplete and that a checklist for stock assessments would be helpful to fishery managers. The committee's checklist is shown in Appendix D. It includes five parts: (1) stock definition, (2) choice of data collection procedures and actual data collection, (3) choice of an assessment model and its parameters, (4) evaluation of alternative hypotheses and possible actions and specification of performance indicators, and (5) presentation of results.

The committee believes it is important for assessments to continue to be conducted by individuals most familiar with the biology of managed species and the associated fishery (i.e., scientists of the federal government and interstate and international fishery management bodies). The National Marine Fisheries Service (NMFS) should be responsible for supporting the long-term collection of fishery-independent data, using either the National Oceanic and Atmospheric Administration (NOAA) fleet or calibrated independent vessels. Diminishing the quality of fishery-independent data by failing to modernize NOAA fishery research vessels or changing sampling methods and gear without proper calibration could imperil existing and future data sets (ASMFC, 1997). The committee did not evaluate the proficiency with which fishery managers communicate stock assessment results and methods to the public. Greater efforts to clarify the workings of stock assessment could be important to increase the credibility of those assessments.

Recommendation: Stock assessment scientists should conduct complete assessments using a checklist such as given in Appendix D. Scientists from state and federal governments and from the independent fisheries commissions should continue to conduct fish stock assessments, with periodic peer review, as described below.

DATA COLLECTION AND ASSESSMENT METHODS

Abundance Indices

Finding: Having an index that is proportional to abundance (called survey index in the committee's simulation) resulted in reasonably good fits of the models. Abundance indices subject to biases or other perturbations (called fishery-dependent in the simulation) can result in poor performance of the models. Combinations of indices, if one is poor, does not improve performance of stock assessment models.

The best index of fish abundance is one for which extraneous influences (e.g., changes in gear, learning, changes in seasonal coverage) can be controlled. Catch per unit effort (CPUE) can vary over time in commercial and recreational fisheries, is subject to fishers' optimizing behaviors, and is not usually the most appropriate index (Chapters 2 and 3). There are examples, however, in which CPUE is an appropriate index if it is interpreted correctly (Quinn and Collie, 1990; NRC, 1994b; see model SS-P6 in Chapter 5). At present, fishery-independent surveys offer the best opportunity for controlling sampling conditions by maintaining consistent gear, spatial coverage, timing, and survey design. Good indices of abundance must be proportional to actual population values.

Recommendation: At the minimum, at least one reliable abundance index should be available for each stock. Fishery-independent surveys offer the best choice for achieving a reliable index if designed well with respect to location, timing, sampling gear, and other statistical survey design considerations.

Obtain Auxiliary Information

Finding: The committee's simulation study demonstrated that assessments are sensitive to underlying structural features of fish stocks and associated fisheries, such as natural mortality, age selectivity, catch reporting, and variations in these and other quantities. For accurate estimation in stock assessment models, it has long been known that auxiliary information in the form of indices or survey estimates of abundance, model structure information, and information about other population parameters (e.g., natural or fishing mortality) improves assessments.

Model performance varied across data sets, and no single method consistently outperformed the others. The more complex models, which allowed for trends in catchability and/or selectivity parameters, tended to work better when only biased abundance indices were used in the fit. However, sophisticated modeling techniques will not fix poor data. Performance of models became erratic as more variability was introduced to data sets. Priority should always be given to collecting the best data possible, by taking into account the biology of the species being studied, the behavior of the fishing fleets, environmental conditions, and survey sampling design. Application of different assessment models to the same data can help to recognize poor data. Results from such comparisons can be used to direct survey programs to improve data quality and to assess the degree of improvement achieved over time. There has been a tendency within NMFS for regional preference of specific assessment methods, which may hinder the identification of bad data. Instead, models should be chosen based on their performance using the data available for a given stock. Alternative methods to estimate parameters of stock assessment models include properly designed and executed tag-recapture and other survey experiments using modern technology to estimate abundance and mortality; methods designed to incorporate seasonal, spatial, and behavioral observations into assessment models; and the use of stomach content information to provide an index of natural mortality over time.

The committee found that different values of natural mortality were assumed or estimated by different analysts, which would account for some of their differences in performance. Therefore, analysts were asked to repeat their analyses using the true value of natural mortality. Natural mortality rates are often assumed to be constant and known. Bayesian techniques discussed in Chapter 3 and implemented in the Autodifferentiation Model Builder (ADMB) that was applied to the simulated data provide a natural way to incorporate independent information about M into the assessment while allowing for the existing uncertainty. A natural mortality index could be included easily in existing assessment models or used in an approach such as multispecies virtual population analysis (MSVPA) that estimates parameters for several species simultaneously.

Although there have been many criticisms of the single-species assessment approach, the committee believes that single-species assessment offers the best approach at present for assessing population parameters and providing short-term forecasting and management advice. It is important to distinguish between such single-species *assessment* procedures and multispecies *management*; the latter topic is beyond the scope of this study. Another form of auxiliary data that could improve stock assessments are measurements of environmental conditions. It is not clear how such data (e.g., spawner-recruit relationships) could be incorporated in stock assessment methods, but there is increasing evidence that environmental conditions are a major factor driving changes in recruitment to some fisheries and that should be considered in developing sampling strategies.

Recommendation: Because there are often problems with the data used in assessments, a variety of different assessment models should be applied to the same data; new methods may have to be developed to evaluate the results of such procedures. The different views provided by different models should improve the quality of assessment results. Greater attention should also be devoted to including independent estimates of natural mortality in assessment models.

More Realistic Assessments of Uncertainty

Finding: Fish stock assessment has often focused on obtaining point estimates (i.e., without uncertainty limits) of the key parameters of the biological system. Management has not tended to follow precautionary approaches.

Precautionary principles established in recent UN agreements call for more emphasis on protection and sustainability of abundance levels and less emphasis on achievement of high levels of catches (FAO, 1995a). The role of stock assessment under these new guidelines is similar to its traditional role, except for the addition of new calculations of the consequences of alternative management actions (see Appendix J). Precautionary fishery management as described by the United Nations Food and Agriculture Organization (FAO, 1995a) requires "taking account of the uncertainties in fisheries systems" (p. 2). According to FAO (1995a):

> The precautionary approach is made more effective by development of an understanding of the sources of uncertainty in the data sampling process, and collection of sufficient information to quantify this uncertainty. If such information is available it can be explicitly used in the management procedure to estimate uncertainty and risk. If such information is not available, a precautionary approach to fishery management would implicitly account for uncertainty by being more conservative. (p. 13)

> A precautionary approach specifically requires a more comprehensive treatment of uncertainty than is the current norm in fishery assessment. This requires recognition of gaps in knowledge, and the explicit identification of the range of interpretations that is reasonable given the present information. (p. 14) [*]

In particular, assessment scientists should provide probability estimates for various consequences, for example, the probability that the stock will exceed some predetermined benchmark levels of abundance 20 years in the future. The need to evaluate management consequences in probabilistic terms will change the way some assessments are conducted. The new types of calculations required can be approached using Bayesian methods; research on other methods (e.g., the applicability of bootstrap Monte Carlo methods to posterior probability calculations) may provide new tools (see Chapter 4). Fishery managers and assessment scientists will have to

---

[*]These principles are not in the form of an international treaty and thus are guidelines rather than binding agreements.

cooperate to fulfill the requirements of the precautionary principle and to implement better feedback procedures. Calculation of the probability of future events requires specification of the rules that management will follow to establish regulations in the future. Because future management rules are difficult to predict, a practical solution is for managers and scientists to consider a range of rules and consequences for which probabilities can be assigned.

The rules that managers follow influence the relative value of surveys and expenditures for other types of applied fisheries research. If a given population is at or near unexploited levels and rules are very cautious in the sense that catches are a small proportion of the population, only minimal survey monitoring may be necessary. Alternatively, heavily exploited populations with aggressive management rules require extensive survey monitoring.

The committee's simulation study demonstrated that small, but realistic, departures from standard assumptions of stock assessment models can result in substantial assessment errors. All too frequently, stock assessments provide only point estimates of key population and management parameters without standard errors or confidence intervals. For the most part, this was the case in the simulation study. The reports from most analysts did not contain measures of uncertainty (the committee did not specifically request them, and the amount of time to do the simulation study was quite limited).

Even if measures of uncertainty are given, they are likely to understate the true level of variation, because uncertainty cannot be included in all parameters. Only recently have posterior probability density functions for key parameters been presented in some stock assessment documents. In our simulation study, only the Kalman filter and ADMB approaches could handle both measurement and process error, both of which are certainly present. Only the most complex models can come close to capturing the uncertainty involved, but it remains unresolved whether the performance of complex models is superior to simpler, possibly more parsimonious, models. The committee evaluated uncertainty primarily by performing retrospective analysis. This analysis showed that persistent under- or overestimation can occur over a number of years of assessment, regardless of which model is used. A risk assessment framework has been adopted in many stock evaluations conducted by NMFS (see Rosenberg and Restrepo, 1994). More broadly, recent symposium volumes attest to the recognition of the importance of this issue in stock assessment (Shepherd, 1991; Smith et al., 1993).

Recommendation: The committee recommends that fish stock assessments include realistic measures of the uncertainty in the output variables whenever feasible. Although a simple model can be a useful management tool, more complex models are needed to better quantify all the unknown aspects of the system and to address the long-term consequences of specific decision rules adequately. The implementation of this recommendation could follow the methods discussed in Chapter 3.

HARVEST STRATEGIES

Finding: Harvest strategies and stock assessments are linked inextricably, although this is not always acknowledged. Management procedures in which the allowable catch is set as a constant fraction of biomass perform better than many alternative procedures, although management based on constant fractional harvests could result in substantial reduction in harvests if biomass is highly uncertain.

Even if the rate of natural mortality assumed in the assessment is correct, uncertainty, correlation in the errors made in successive years, and persistent bias in the estimation of stock abundance can significantly degrade management performance. The simulation results reported herein, albeit limited, illustrate that assessment errors can be substantial, even when unbiased abundance indices are available. Thus, harvesting policies that are robust to these levels of errors are necessary.

Instead of harvesting a constant fraction of exploitable biomass, fishing mortality could be reduced sequentially at low stock sizes or a threshold biomass level could be established below which fishing would be curtailed (which is done for many North Pacific species). In addition to lowering target harvest levels, alternative controls may be needed as a safeguard against major assessment errors. Increasing the size limits with gear restrictions could allow a higher target fishing mortality rate (F) to be applied with less risk. Area closures may be appropriate for sedentary species if the distribution of organisms is well known. Area closures can also be appropriate for migratory species, in that they can protect the species from overharvesting during part of the year.

Recommendation: Precautionary management procedures should include management tools specific to the species managed, such as threshold biomass levels, size limits, gear restrictions, and area closures (for sedentary species).

RIGOROUS EVALUATION SYSTEM

Finding: In many U.S. fisheries, biological reference points are used to set target harvest rates. Justification for the use of various reference points comes from analyses of empirical and theoretical evidence encompassing a wide variety of stock categories. The aim of these studies has been to determine reference points that lead to sustainable harvest levels for stocks of a given life-history type. However, the performance of harvesting strategies and regulatory decision rules depend on the performance of the assessment method used to implement them (Chapter 4), which may vary widely among stocks depending on data availability.

The total allowable catch (TAC) for many U.S. West Coast fisheries is specified by applying $F_{40\%}$ to an estimate of exploitable biomass, corresponding to a harvest that reduces the equilibrium spawning biomass per recruit to 40% of what it would be under no harvesting. With this method, if the estimated natural mortality rate is too high, $F_{40\%}$, the estimated exploitable biomass, and TAC will all be too high (Table 5.3, compare data sets 1 and 2). Even if the correct natural mortality rate is known, biases in the stock assessment, uncertainty, or correlated errors in the biomass projection can degrade management performance. FAO (1995a, pp. 14-15) noted:

> A precautionary approach to assessment and analysis requires a realistic appraisal of the range of outcomes possible under fishing and the chances of these outcomes under different management actions. The precautionary approach to assessment would follow a process of identifying alternative possible hypotheses of states of nature, based on the information available, and examining the consequences of proposed management actions under each of these alternative hypotheses. This process would be the same in data rich and data poor analyses. A precautionary assessment would, at least, aim to consider (a) uncertainties in data; (b) specific alternative hypotheses about underlying biological, economic and social processes, and (c) calculation of the response of the system to a range of alternative management actions.

Recommendation: Assessment methods and harvesting strategies have to be evaluated simultaneously to determine their ability to achieve management goals. Ideally, this involves implementing them both in simulations of future stock trajectories. For complex assessment methods, this may prove to be very computationally intensive, and an alternative is to simulate only the decision rule while making realistic assumptions about the uncertainty of future assessments. Simulation models should be realistic and should encompass a wide range of possible stock responses to management and natural fluctuations consistent with historical experience. The performance of alternative methods and decision rules should be evaluated using several criteria, including the distribution of yield and the probabilities of exceeding management thresholds.

NEW APPROACHES

Finding: Stock assessments are compromised by incomplete or variable data and new methods will be needed to deal with this situation.

The committee was reluctant to create a prescriptive list of research needs for improving stocks assessments because it is difficult to foresee where the science and practice of stock assessments should proceed in the future and because such a list could hinder research in directions not included on it. The most important action that should be taken by NMFS and other organizations that depend on stock assessments would be to create and maintain an environment that fosters the development of a range of new techniques.

A few prominent themes emerged from the committee's discussions. The use of Bayesian methods and other means of incorporating uncertainty in stock assessment models should be pursued aggressively. One committee member believes that interval analysis, fuzzy arithmetic, and fuzzy logic may be preferable in some instances to Bayesian approaches and that the former techniques should be investigated for possible application. Assessment models for recreational fisheries are often less developed than models for commercial fisheries and require further development. Catchability, selectivity, and mortality are often assumed to be constant over time. When these factors are assumed to be constant but are not, it can lead to faulty stock assessments. New means to estimate changes in catchability, selectivity, and mortality over time should be developed along with models to include such data.

The committee's simulations showed that the biomass of recovering populations tends to be underestimated when survey data are used. New means of accounting for stock recovery must be developed. Although few stocks seem to be in recovery phases, NMFS and the fishery councils are required by law to develop recovery plans for depleted stocks. It is important to track stock recovery accurately to minimize economic disruption in a fishery while protecting the stock from depletion in the future.

Recommendation: NMFS and other bodies responsible for fishery management should support the development of new techniques for stock assessment that are robust to incomplete, ambiguous, and variable data and to the effects of environmental fluctuations on fisheries.

PEER REVIEW OF ASSESSMENTS AND ASSESSMENT METHODS

Finding: External peer review of scientific procedures and results is standard practice throughout much of the scientific community. When applied properly to stock assessments, such reviews would yield an impartial evaluation of the quality of assessments as well as constructive suggestions for improvement. These reviews are most beneficial when conducted periodically, for example, every 5 to 10 years, as new information and practices develop.

Stock assessments are often the focus of disputes among fishery managers, fishers, and environmental groups. It is imperative that stock assessment procedures be understood better and trusted more by all stakeholders. FAO (1995a, p. 14) focused on this issue:

> Specifically, the assessment process should include:
>
> a. scientific standards of evidence (objective, verifiable, and potentially replicable), should be applied in the evaluation of information used in analysis;
> b. a process for assessment and analysis that is transparent; and
> c. periodic, independent, objective, and in-depth peer review as a quality assurance.

Recommendation: The committee recommends that NMFS conduct (at reasonable intervals) in-depth, independent peer review of its fishery management methods to include (1) the survey sampling methods used in the collection of fishery and fishery-independent data, (2) stock assessment procedures, and (3) management and risk assessment strategies.

Commercial Data Collection Protocols

Finding: The committee found that formal sampling protocols for collection of commercial fisheries statistics were unavailable for some geographic regions.

Some regions, for example, the U.S. Northeast, are in the midst of the development and publication of new protocols, whereas other regions do not appear to use standardized methods. The lack of formalized, peer-reviewed data collection methods in commercial fisheries is worrisome. To the extent that formalized and standardized procedures are lacking, potential bias and improper survey conduct may exist, with unknown impact on data reliability. The committee's simulations show that 30% underreporting of fisheries data diminishes the accuracy of stock assessments. Other than this result, there has been little study of the potential impact of unreliable sampling practices on assessment accuracy, and the magnitude of this problem is unknown. Formalized sampling protocols have been developed for recreational fisheries in the form of the Marine Recreational Fisheries Statistics Survey (MRFSS). MRFSS data and methods have undergone independent peer review, are readily available, and could serve as a model for commercial fisheries. MRFSS data and methods are not perfect, however. The MRFSS design does not provide precise estimates for the types of angling that require a specialized and targeted survey, such as for highly migratory species, charter boat fisheries, or species subject to a short fishing season.

Recommendations: The committee recommends that a standardized and formalized data collection protocol be established for commercial fisheries data nationwide. The committee further recommends that a complete review of methods for collection of data from commercial fisheries be conducted by an independent panel of experts.

EDUCATION AND TRAINING

Finding: Reduction in the supply of stock assessment scientists would endanger the conduct of fishery assessments by the federal government, interstate commissions, and international management bodies and would hinder progress in the development and implementation of new stock assessment methods.

The training of stock assessment scientists should endow them with skills in applied mathematics, fisheries biology, and oceanography. Such training should begin at the undergraduate level and continue throughout a scientist's professional career. Education of fisheries scientists can and should be organized in such a way that it complements and augments the NMFS research mission and leads to improved management strategies for fisheries. Skills needed include a knowledge of the biology of the target species and of the associated fisheries, as well as applied mathematics, including probability, statistics (maximum likelihood theory, nonlinear functions, statistical decision theory, Bayesian methods, survey sampling theory), and modeling. Training can be accomplished through a variety of mechanisms, including (1) coursework as part of graduate studies; (2) internships with experienced stock assessment professionals; (3) in-service training courses offered on-site; (4) professional leave to take university courses; (5) personnel exchanges of stock assessment scientists working in academia and government laboratories; and/or (6) time spent observing actual fishing or survey operations. It is commendable that NOAA has established some formal programs to foster education and collaborative exchange with academic scientists. Examples are the NOAA-University of Miami Cooperative Institute for Marine and Atmospheric Studies in the southeast and the NOAA-multiuniversity Cooperative Marine Education and Research program in the northeast.

NMFS should support undergraduate participation in research to advance stock assessment research while providing educational opportunities that interest undergraduates in fishery science as a career. Such opportunities might include research internships for undergraduate juniors and seniors in federal or university fisheries laboratories. The Research Experience for Undergraduates program funded by the National Science Foundation could serve as a model. The committee reiterates the recommendation of the National Research Council (NRC) Ocean Studies Board that "federal agencies with marine-related missions [should] find mechanisms to guarantee the continuing vitality of the underlying basic science on which they depend" (NRC, 1992, p. 10).

Recommendations: NMFS and other bodies that conduct fish stock assessments should ensure a steady supply of well-trained stock assessment scientists to conduct actual assessments and to carry out associated research. NMFS should encourage partnerships among universities, government laboratories, and industry for their mutual benefit. This can be accomplished by exchanging personnel and ideas and by providing funding for continuing education at the graduate, postdoctoral, and professional levels, including elements such as cooperative research projects and specialized courses, workshops, and symposia.

References

Adams, D.C., J. Gurevitch, and M.S. Rosenberg. 1997. Resampling tests for meta-analysis of ecological data. *Ecology* 78:1277-1283.

Aglen, A., and O. Nakken. 1994. Length dependent corrections of survey estimates of cod and haddock in the Barents Sea. *ICES C.M.* 1994/ G.

Alefield, G., and J. Herzberger. 1983. *Introduction to Interval Computations.* Academic Press, New York.

Alverson, D.L., M.H. Freeberg, S.A. Murawski, and J.G. Pope. 1994. *A Global Assessment of Fisheries Bycatch and Discards.* FAO Technical Paper 339. Rome.

Anderson, R.O., and H.M. LeRoy. 1969. Angling as a factor influencing catchability of largemouth bass. *Trans. Amer. Fish. Soc.* 98:317-320.

Anonymous. 1995a. *Report of the Working Group on Methods of Fish Stock Assessment.* ICES Cooperative Research Report 199. Copenhagen.

Anonymous. 1995b. *Incorporating Uncertainty Into Stock Projections: Report of the Scientific Meeting.* April 3-7. Hobart, Tasmania, CCSBT Document SBFWS/95/inf-3.

Argue, A.W., R. Hilborn, R.M. Peterman, M.J. Staley, and C.J. Walters. 1983. The strait of Georgia chinook and coho fishery. *Can. Bull. Fish. Aquat. Sci.* 211.

Arnason, R. 1993. ITQ based fisheries management. Pp. 345-356 in S.J. Smith, J.J. Hunt, and D. Rivard (eds.), *Risk Evaluation and Biological Reference Points for Fisheries Management.* Canadian Special Publication of Fisheries and Aquatic Sciences 120.

Atlantic States Marine Fisheries Commission (ASMFC). 1997. *Proceedings of the Workshop on Maintaining Current and Future Fisheries Resource Survey Capabilities.* Special Report No. 63. ASMFC, Washington, D.C.

Atwood, C.G., and B.A. Bennett. 1990. A simulation model of the sport fishery for Galjoen *Coracinus capensis*: An evaluation of minimum size limit. *S. Afr. J. Mar. Sci.* 9:359-369.

Atwood, C.G., and B.A. Bennett. 1995. Modeling the effect of marine reserves on the recreational shore-fishery of the South-Western Cape, South Africa. *S. Afr. J. Mar. Sci.* 16:227-240.

Azarovitz, T.R. 1981. A brief historical review of the Woods Hole trawl survey time series. Pp. 62-67 in W.G. Doubleday and D. Rivard (eds.), *Bottom Trawl Surveys.* Canadian Special Publication of Fisheries and Aquatic Sciences 58.

Baird, J.W., and S.C. Stevenson. 1983. Levels of precision—Sea versus shore sampling. Pp. 185-188 in W.G. Doubleday and D. Rivard (eds.), *Sampling Commercial Catches of Marine Fish and Invertebrates.* Canadian Special Publication of Fisheries and Aquatic Sciences 66.

Beamish, R.J., and D.R. Bouillon. 1993. Pacific salmon production trends in relation to climate. *Can. J. Fish. Aquat. Sci.* 50:1002-1016.

Beddington, J.R., and J. Cooke. 1983. *The Potential Yield of Previously Unexploited Stocks.* FAO No. 242. Food and Agriculture Organization, Fisheries Technical Paper, United Nations. Rome.

Beddington, J.R., and R.M. May. 1977. Harvesting natural populations in a randomly fluctuating environment. *Science* 197:463-465.

Bence, J.R., A. Gordoa, and J.E. Hightower. 1993. Influence of age-selective surveys on the reliability of Stock Synthesis assessments. *Can. J. Fish. Aquat. Sci.* 50:827-840.

Beverton, R.J.H., and S.J. Holt. 1993. *On the Dynamics of Exploited Fish Populations.* Chapman and Hall, London.

Borgstrom, R. 1992. Effect of population density on gillnet catchability in four allopatric populations of brown trout (*Salmo trutta*). *Can. J. Fish. Aquat. Sci.* 49:1539-1545.

Bowman, R.E., and E.W. Bowman. 1980. Diurnal variation in the feeding intensity and catchability of silver hake (*Merluccius bilinearis*). *Can. J. Fish. Aquat. Sci.* 49:1539-1545.

Brauhn, J L., and H. Kincaid. 1982. Survival, growth, and catchability of rainbow trout of four strains. *N. Am. J. Fish. Manage.* 2:1-10.

Brown, T.L. 1991. Use and abuse of mail surveys in fisheries management. *Amer. Fish. Soc. Symp.* 12:255-261.

Buijse, A.D., L.A. Schaap, and T.P. Bult. 1992. Influence of water clarity on the catchability of six freshwater fish species in bottom trawls. *Can. J. Fish. Aquat. Sci.* 49:885-893.

Burns, T.S., R. Schultz, and B.E. Brown. 1983. The commercial catch sampling program in the northeastern United States. Pp. 82-95 in W.G. Doubleday and D. Rivard (eds.), *Sampling Commercial Catches of Marine Fish and Invertebrates.* Canadian Special Publication of Fisheries and Aquatic Sciences 66.

Buxton, C.A. 1992. The application of yield-per-recruit models to two South African sparid reef species, with special consideration of sex change. *Fish. Res.* 15:1-16.

Carl, L.M., B.J. Shuter, and J.E. Matuszek. 1991. Production and instantaneous growth rates of lake trout before and after lake herring introduction in Lake Opeongo, Ontario. *Amer. Fish. Soc. Symp.* 12:476-480.

Clark, C.W. 1985. *Bioeconomic Modelling and Fisheries Management.* John Wiley & Sons, New York.

Clark, W.G. 1991. Groundfish exploitation rates based on life history parameters. *Can. J. Fish. Aquat. Sci.* 48:734-750

Clark, W.G. 1993. The effect of recruitment variability on the choice of a target level of spawning biomass per recruit. Pp. 233-246 in G. Kruse, D.M. Eggers, R.J. Marasco, C. Pautzke, and T.J. Quinn II (eds.), *Proceedings of the International Symposium on Management Strategies for Exploited Fish Populations.* Alaska Sea Grant College Program Report No. 93-02. University of Alaska, Fairbanks.

Clark, W.G. 1996. Long-term changes in halibut size at age. Pp. 55-62 in *Report of Assessment and Research Activities, International Pacific Halibut Commission—1995.* International Pacific Halibut Commission, Seattle, Wash.

Claytor, R.R., E.M.P. Chadwick, G.A. Nielsen, G.J. Chaput, D.K. Cairns, S.C. Courtenay, and H.M.C. Dupuis. 1991. Index programs: Their value in southern Gulf of St. Lawrence fish stock assessment. *ICES C.M.* D:1.

Cochran, W.G. 1977. *Sampling Techniques.* John Wiley & Sons, New York.

Collie, J.S., and G.H. Kruse. 1997. Estimating king crab abundance from commercial catch and research survey data. Canadian Special Publication of Fisheries Aquatic Sciences 125. In press.

Collie, J.S., and P.D. Spencer. 1993. Management strategies for fish populations subject to long-term environmental variability and depensatory predation. Pp. 629-650 in G. Kruse, D.M. Eggers, R.J. Marasco, C. Pautzke, and T.J. Quinn II (eds.), *Proceedings of the International Symposium on Management Strategies for Exploited Fish Populations.* Alaska Sea Grant College Program Report No. 93-02. University of Alaska, Fairbanks.

Collins, J.J. 1987. Increased catchability of the deep monofilament nylon gillnet and its expression in a simulated fishery. *Can. J. Fish. Aquat. Sci.* 44:129-135.

Conan, G.Y., and E. Wade. 1989. Geostatistical mapping and global estimation of harvestable resources in a fishery of northern shrimp (*Pandalus borealis*). International Council for the Exploration of the Sea C.M. 1989/D:1.

Conser, R., 1993. A brief history of ADAPT. *Northw. Atl. Fish. Org., Sci. Council Studies* 17:83-87.

Conser, R., and J.E. Powers. 1990. Extensions of the ADAPT VPA tuning method designed to facilitate assessment work on tuna and swordfish stocks. ICCAT Working Document SCRS/89/43.

Crecco, V., and W.J. Overholtz. 1990. Causes of density-dependent catchability for Georges Bank haddock *Melanogrammus aeglefinus*. *Can. J. Fish. Aquat. Sci.* 47:385-394.

Crecco, V., and T. Savoy. 1985. Density-dependent catchability and its potential causes and consequences on Connecticut River American shad, *Alosa sapidissima. Can. J. Fish. Aquat. Sci.* 42:1649-1657.

Cressie, N.A.C. 1993. *Statistics for Spatial Data.* Revised edition. John Wiley & Sons. New York.

Crone, P.R. 1995. Sampling design and statistical considerations for the commercial groundfish fishery of Oregon. *Can. J. Fish. Aquat. Sci.* 52:716-732.

Cushing, D.H., and R.R. Dickson. 1976. The biological response in the sea to climatic changes. *Adv. Mar. Biol.* 14:1-122.

D'Agostino, R.B., and M. Weintraub. 1995. Meta-analysis: A method for synthesizing research. *Clin. Pharmacol. Ther.* 58:605-616.

D'Amours, D. 1993. The distribution of cod (*Gadus morhua*) in relation to temperature and oxygen level in the Gulf of St. Lawrence. *Fish. Oceanogr.* 2:24-29.

de Lafontaine, Y., T. Lambert, G.R. Lilly, W.D. McKone, and R.J. Miller (ed.). 1991. Juvenile Stages: The Missing Link in Fisheries Research. Report of a Workshop. *Canadian Technical Report in Fisheries and Aquatic Sciences* 1890.

Dennis, B. 1996. Should ecologists become Bayesians? *Ecol. Appl.* 6:1095-1103.

Deriso, R.B. 1980. Harvesting strategies and parameter estimation for an age-structured model. *Can. J. Fish. Aquat. Sci.* 37:268-282.

Deriso, R.B. 1985. Risk averse harvesting strategies. Pp. 65-72 in M. Mangel (ed.), *Resource Management.* Springer-Verlag, Berlin.

Deriso, R.D. 1987. Optimal $F_{0.1}$ criteria and their relationship to maximum sustainable yield. *Can. J. Fish. Aquatic. Sci.* 44(Suppl. 2):339-348.

Deriso, R.B., and A.M. Parma. 1988. Dynamics of age and size for a stochastic population model. *Can. J. Fish. Aquat. Sci.* 45:1054-1068.

Deriso, R.B., T.J. Quinn II, and P.R. Neal. 1985. Catch-age analysis with auxiliary information. *Can. J. Fish. Aquat. Sci.* 42:815-824.

Dickson, W. 1993a. Estimation of the capture efficiency of trawl gear. I: Development of a theoretical model. *Fish. Res.* 16:239-253.

Dickson, W. 1993b. Estimation of the capture efficiency of trawl gear. II: Testing a theoretical model. *Fish. Res.* 16:255-272.

Dillman, D.A. 1978. *Mail and Telephone Surveys: The Total Design Method.* John Wiley & Sons, New York.

Doubleday, W.G. 1976. A least squares approach to analyzing catch at age data. *ICNAF Res. Bull.* 12:69-81.

Doubleday, W.G., and D. Rivald. 1981. *Bottom Trawl Surveys.* Canadian Special Publication of Aquatic Sciences 58.

Dubols, D., and H. Prade. 1988. *Possibility Theory: An Approach to Computerized Processing of Uncertainty.* Plenum Press, New York.

Dwyer, W.P. 1990. Catchability of three strains of cutthroat trout. *N. Am. J. Fish. Manage.* 10:458-461.

Ecker, M.D., and J.F. Heltshe. 1994. Geostatistical estimates of scallop abundance. Pp. 125-144 in N. Lange, L. Ryan, L. Billard, D. Brillinger, L. Conquest, and J. Greenhouse (eds.), *Case Studies in Biometry.* John Wiley & Sons, New York.

Eddy, D.M., V. Hasselblad, and R. Schachter. 1992. *Meta-analysis by the Confidence Profile Method.* Academic Press, New York.

Eggers, D.M. 1993. Robust harvest policies for Pacific salmon fisheries. Pp. 85-106 in G. Kruse, D.M. Eggers, R.J. Marasco, C. Pautzke, and T.J. Quinn II (eds.), *Proceedings of the International Symposium on Management Strategies for Exploited Fish Populations.* Alaska Sea Grant College Program Report No. 93-02. University of Alaska, Fairbanks.

Engås, A., and A.V. Soldal. 1992. Diurnal variations in bottom trawl catch rates of cod and haddock and their influence on abundance indices. *ICES J. Mar. Sci.* 49:89-95.

Engstrom-Heg, R. 1986. Interaction of area with catchability indices used in analyzing inland recreational fisheries. *Trans. Amer. Fish. Soc.* 115:818-822.

Essig, R.J., and M.C. Holliday. 1991. Development of a recreational fishing survey: The marine recreational fisheries statistics survey case study. *Amer. Fish. Soc. Symp.* 12:245-254.

Fabrizio, M.C., and R.A. Richards. 1996. Commercial fisheries surveys. Pp. 625-650 in B.R. Murphy and D.W. Willis (eds.), *Fisheries Techniques*, 2nd edition. American Fisheries Society, Bethesda, Md.

Ferson, S., and L.R. Ginzburg. In press. Different methods are needed to propagate ignorances and variability. *Reliability Engineering and System Safety.*

Ferson, S. 1994. Using fuzzy arithmetic in Monte Carlo simulation of fishery populations. Pp. 595-608 in G. Kruse, D.M. Eggers, R.J. Marasco, C. Pautzke, and T.J. Quinn II (eds.), *Proceedings of the International Symposium on Management Strategies for Exploited Fish Populations.* Alaska Sea Grant College Program Report No. 93-02. University of Alaska, Fairbanks.

Ferson. S. 1995. Quality assurance for Monte Carlo risk assessment. Pp. 14-19 in B.M. Ayyub (ed.), *Proceedings of ISUMA-NAFIPS'96.* IEEE Computer Society Press, Los Alamitos, Calif.

Ferson, S., and R. Kuhn. 1994. RiskCalc^{TM}: *Uncertainty Analysis with Interval and Fuzzy Arithmetic.* Applied Biomathematics, Setauket, N.Y.

Fletcher, R.I. 1978. Time-dependent solutions and efficient parameters for stock-production models. *Fish. Bull.* 76:377-388.

Food and Agriculture Organization (FAO). 1995a. *Precautionary Approach to Fisheries.* FAO Fisheries Technical Report 350. United Nations, Rome.

Food and Agriculture Organization (FAO). 1995b. *The State of World Fisheries and Agriculture.* United Nations, Rome.

Fournier, D.A., and C.P. Archibald. 1982. A general theory for analyzing catch at age data. *Can. J. Fish. Aquat. Sci.* 39:1195-1207.

Fournier, D.A., and I.J. Doonan. 1987. A length-based stock assessment method utilizing a generalized delay-difference model. *Can. J. Fish. Aquat. Sci.* 44:422-437.

Francis, R.I.C.C. 1984. An adaptive strategy for stratified random trawl surveys. *NZ. J. Mar. Res.* 18:59-71.

Frederick, S.W., and R.M. Peterman. 1995. Choosing fisheries harvest policies: When does uncertainty matter? *Can. J. Fish. Aquat. Sci.* 52:291-306.

Frey, J.H. 1983. *Survey Research by Telephone.* Sage Publications, Beverly Hills, Calif.

Fry, F.J. 1949. Statistics of a lake trout fishery. *Biometrics* 5:27-67.

Gavaris, S. 1988. An adaptive framework for the estimation of population size. CAFSAC Research Document 88129.

Gavaris, S. 1993. Analytical estimates of reliability for the projected yield from commercial fisheries. Pp. 185-191 in S.J. Smith, J.J. Hunt, and D. Rivard (eds.), *Risk Evaluation and Biological Reference Points for Fisheries Management.* Canadian Special Publication of Fisheries and Aquatic Sciences 120.

Geiger, H.J., and J.P. Koenings. 1991. Escapement goals for sockeye salmon with informative prior probabilities based on habitat considerations. *Fish. Res.* 11:239-256.

Gelman, A., J.B. Carlin, H.S. Stern, and D.B. Rubin. 1995. *Bayesian Data Analysis.* Chapman and Hall, London.

Getz, W.M., and R.G. Haight. 1989. *Population Harvesting: Demographic Models of Fish, Forest, and Animal Resources.* Princeton University Press, Princeton, N.J.

Godø, O.R. 1994. Natural fish behaviour and catchability of groundfish. *ICES C.M.* G:14.

Godø, O.R., and A. Engås. 1989. Swept area variation with depth and its influence on abundance indices of groundfish from trawl surveys. *J. Northwest Atl. Fish. Sci.* 9:133-139.

Godø, O.R., and K. Sunnanå. 1992. Size selection during trawl sampling of cod and haddock and its effect on abundance indices at age. *Fish. Res.* 13:231-239.

Godø, O.R., and V.G. Wespestad. 1993. Monitoring changes in abundance of gadoids with varying availability to trawl and acoustic surveys. *ICES J. Mar. Sci.* 50:39-52.

Gordoa, A., and J.E. Hightower. 1991. Changes in catchability in a bottom-trawl fishery for Cape hake (*Merluccius capensis*). *Can. J. Fish. Aquat. Sci.* 48:1887-1895.

Groves, R.M. 1989. *Survey Errors and Survey Costs.* John Wiley & Sons, New York.

Groves, R.M., P.P. Biemer, L.E. Lyberg, J.T. Massey, W.L. Nicholls, II, and J. Waksberg (eds.). 1988. *Telephone Survey Methodology.* John Wiley & Sons, New York.

Guillard, J., D. Gerdeaux, G. Brun, and R. Chappez. 1992. The use of geostatistics to analyze data from an echo-integration survey of fish stock in Lake Sainte-Croix. *Fish. Res.* 13:395-406.

Gulland, J.A. 1966. *Manual of Sampling and Statistical Methods for Fisheries Biology.* FAO Manual of Fisheries Science 3. U.N. Food and Agriculture Organization, Rome.

Gulland, J.A. 1969. *Manual of Methods of Fish Stock Assessment.* FAO Manual of Fisheries Science 4. U.N. Food and Agriculture Organization, Rome.

Gulland, J.A., and L.K. Boerma. 1973. Scientific advice on catch levels. *Fish. Bull.* 71:325-335.

Gunderson, D.R. 1993. *Surveys of Fisheries Resources.* John Wiley & Sons, New York.

Hall, D.L., R. Hilborn, M. Stocker, and C.J. Walters. 1988. Alternative harvest strategies for Pacific herring (*Clupea harengus pallasi*). *Can. J. Fish. Aquat. Sci.* 45:888-897.

Hannah, R.W. 1995. Variation in geographic stock area, catchability, and natural mortality of ocean shrimp (*Pandalus jordani*): Some evidence for a trophic interaction with Pacific hake (*Merluccius productus*). *Can. J. Fish. Aquat. Sci.* 52:1018-1029.

Hanneson, R. 1989. Fixed or variable catch quotas? The importance of population dynamics and stock dependent costs. Pp. 459-465 in P.A. Neher, R. Arnason, and N. Mollet (eds.), *Rights-Based Fishing.* Kluwer Academic Publishers, Dordrecht, Netherlands.

Hayne, D.W. 1991. The access point survey: Procedures and comparison with the roving-clerk creel survey. *Amer. Fish. Soc. Symp.* 12:123-138.

He, P. 1991. Swimming endurance of the Atlantic cod, *Gadus morhua* L., at low temperatures. *Fish. Res.* 12:65-73.

Hedges, L.Y., and I. Olkin. 1985. *Statistical Methods for Meta-Analysis.* Academic Press, San Diego, Calif.

Helser, T.E., and D.B. Hayes. 1995. Providing quantitative management advice from stock abundance indices based on research surveys. *Fish. Bull.* 93:290-298.

Hilborn, R. 1979. Comparison of fisheries control systems that utilize catch and effort data. *J. Fish. Res. Bd. Can.* 36:1477-1489.

Hilborn, R., and C.J. Walters. 1992. *Quantitative Fisheries Stock Assessment: Choice Dynamics and Uncertainty.* Routledge, Chapman & Hall, New York.

Hilborn, R., E.K. Pikitch, and M.K. McAllister. 1994. A Bayesian estimation and decision analysis for an age-structured model using biomass survey data. *Fish. Res.* 19:17-30.

Hobert, J.P., and G. Casella. 1996. The effect of improper priors on Gibbs sampling in hierarchical linear mixed models. *J. Amer. Stat. Assoc.* 91:1461-1473.

Hoenig, J.M., D.S. Robson, C.M. Jones, and K.H. Pollock. 1993. Scheduling counts in the instantaneous and progressive count methods for estimating sport fishing effort. *N. Am. J. Fish. Manage.* 13:723-736.

Horbowy, J. 1992. The differential alternative to the Deriso difference production model. *ICES J. Mar. Sci.* 49:167-174.

Hutchings, J., and R.A. Myers. 1994. What can be learned from the collapse of a renewable resource? Atlantic cod, *Gadus morhua*, of Newfoundland and Labrador. *Can. J. Fish. Aquat. Sci.* 51:2126-2146.

Inman, C.R., R.C. Dewey, and P.P. Durocher. 1977. Growth comparisons and catchability of three largemouth bass strains. *Fisheries* 2:20-25.

International Council for the Exploration of the Sea (ICES). 1993. *Report of the Working Group on Methods of Fish Stock Assessments.* ICES Cooperative Research Report No. 191. Copenhagen.

Jinn, J.H., J. Sedransk, and P.J. Smith. 1987. Optimal two-phase stratified sampling for estimation of the age composition of a fish population. *Biometrics* 43:343-353.

Jolly, G.M., and I. Hampton. 1990. A stratified random transect design for acoustic surveys of fish stocks. *Can. J. Fish. Aquat. Sci.* 47:1282-1291.

Jones, C.M., and D.S. Robson. 1991. Improving precision in recreational angling surveys: The traditional access method versus the busroute design. *Amer. Fish. Soc. Symp.* 12:177-188.

Jones, C.M., D.S. Robson, D. Otis, and S. Gloss. 1990. Use of a computer simulation model to determine the behavior of a new survey estimator of recreational angling. *Trans. Amer. Fish. Soc.* 119:41-54.

Jones, C.M., D.S. Robson, H.D. Lakkis, and J. Kressel. 1995. Properties of catch rates used in analysis of angler surveys. *Trans. Amer. Fish. Soc.* 124:911-928.

Kass, R.E., and L. Wasserman. 1996. The selection of priors by formal rules. *J. Amer. Stat. Assoc.* 91:1343-1370.

Kaufmann, A., and M.M. Gupta. 1985. *Introduction to Fuzzy Arithmetic: Theory and Applications.* Van Nostrand Reinhold, New York.

Kerr, S. 1992. Artificial Reefs in Australia. Their Construction, Location, and Function. Working Papers. Bureau of Rural Resources, Canberra, Australia. WP/8/92.

Kimura, D.K. 1990. Approaches to age-structured separable sequential population analysis. *Can. J. Fish. Aquat. Sci.* 47:2364-2374.

Kleinsasser, L.J., W. Holt, and B. Wjiteside. 1990. Growth and catchability of northern, Florida, and F1 hybrid largemouth bass in Texas ponds. *N. Am. J. Fish. Manage.* 10:462-468.

Koeller, P.A. 1991. Approaches to improving groundfish survey abundance estimates by controlling the variability of survey gear geometry and performance. *J. Northwest Atl. Fish. Sci.* 11:51-58.

Korsbrekke, K., S. Mehl, O. Nakken, and K. Sunnana. 1995. Bunnfiskunderskelser I Barentsthavet Vinteren 1995 (Investigations on demersal fish in the Barents Sea Winter 1995). Fisken og Havet (Cruise Report) 13. Institute for Marine Research, Bergen, Norway (Norwegian text and English figure and table captions).

Krieger, K.J., and M.F. Sigler. 1996. Catchability coefficient for rockfish estimated from trawl and submersible surveys. *Fish. Bull.* 94:282-288.

Kulka, D.W., and D. Waldron. 1983. The Atlantic observer programs—A discussion of sampling from commercial catches at sea. Pp. 255-262 in W.G. Doubleday and D. Rivard (eds.), *Sampling Commercial Catches of Marine Fish and Invertebrates.* Canadian Special Publication of Fisheries and Aquatic Sciences 66.

Lepkowski, J.M. 1988. Telephone sampling methods in the United States. Pp. 73-97 in R.M. Groves, P.P. Biemer, L.E. Lyberg, J.T. Massey, W.L. Nicholls II, and J. Waksberg (eds.), *Telephone Survey Methodology.* John Wiley & Sons, New York.

Liermann, M., and R. Hilborn. In press. Hierarchic Bayesian meta analysis of depensation in marine fish stocks. *Can. J. Fish. Aquat. Sci.*

Linhart, H., and W. Zucchini. 1986. *Model Selection.* John Wiley & Sons, New York.

Lluch-Belda D., R.A. Schwartzlose, R. Serra, R. Parrish, T. Kawasaki, D. Hedgecock, and R.J.M. Crawford. 1992. Sardine and anchovy regime fluctuations of abundance in four regions of the world oceans: A workshop report. *Fish. Oceanogr.* 1:339-347.

Loesch, J.G., W.H. Kriete, Jr., and E.J. Foell. 1982. Effects of light intensity on the catchability of juvenile anadramous *Alosa* species. *Trans. Amer. Fish. Soc.* 111: 41-44.

Ludwig, D., and C.J. Walters. 1985. Are age-structured models appropriate for catch-effort data? *Can. J. Fish. Aquat. Sci.* 42:1066-1072.

Ludwig, D., and C.J. Walters. 1989. A robust method for parameter estimation from catch and effort data. *Can. J. Fish. Aquat. Sci.* 46:137-144.

Mace, P. 1994. Relationships between common biological reference points used as thresholds and targets of fisheries management strategies. *Can. J. Fish. Aquat. Sci.* 51:110-122.

Mace, P.M., and M.P. Sissenwine. 1993. How much spawning biomass per recruit is enough? Pp. 101-118 in S.J. Smith, J.J. Hunt, and D. Rivard (eds.). *Risk Evaluation and Biological Reference Points for Fisheries Management.* Canadian Special Publication of Fisheries and Aquatic Sciences 120.

MacLennan, D.N., and E.J. Simmonds. 1992. *Fisheries Acoustics.* Chapman and Hall, London.

Malvestuto, S.P. 1983. Sampling the recreational fishery. Pp. 397-419 in L.A. Nielsen and D.L. Johnson (eds.), *Fisheries Techniques.* American Fisheries Society, Bethesda, Md.

Malvestuto, S.P., W.D. Davies, and W.L. Shelton. 1978. An evaluation of the roving creel survey with nonuniform probability sampling. *Trans. Amer. Fish. Soc.* 107:255-262.

Mangel, M. 1985. *Decision and Control in Uncertain Resource Systems.* Academic Press, New York.

Manly, B.F.J. 1991. *Randomization and Monte-Carlo Methods in Biology.* Chapman and Hall, London.

Marshall, D., O. Johnell, and H. Wedel. 1996. Meta-analysis of how well measures of bone mineral density predict occurrence of osteoporotic fractures. *Brit. Med. J.* 312:1254-1259.

McAllister, M.M., E.K. Pikitch, A.E. Punt, and R. Hilborn. 1994. A Bayesian approach to stock assessment and harvest decisions using the sampling/importance resampling algorithm. *Can. J. Fish. Aquat. Sci.* 51:2673-2687.

Megrey, B.A. 1989. Review and comparison of age-structured stock assessment models. *Amer. Fish. Soc. Symp.* 6:8-48.

Mendelsson, R. 1982. Discount factors and risk aversion in managing random fish populations. *Can. J. Fish. Aquat. Sci.* 39:1252-1257.

Merritt, M.F. 1995. Application of Decision Analysis to the Evaluation of Recreational Fishery Management Problems. Ph.D. dissertation, University of Alaska, Fairbanks.

Methot, R.D. 1989. Synthetic estimates of historical abundance and mortality in northern anchovy. *Amer. Fish. Soc. Symp.* 6:66-82.

Methot, R.D. 1990. Synthesis model: An adaptable framework for analysis of diverse stock assessment data. *International North Pacific Fishery Commission Bull.* 50:259-277.

Michalsen, K., O.R. Godø, and A. Ferro 1996. Diel variation in the catchability of gadoids and its influence on the reliability of abundance indices. *ICES J. Mar. Sci.* 53:389-395.

Miller, A.J., D.R. Cayan, T.P. Barnett, N.E. Graham, and J.M. Oberhuber. 1994. The 1976-77 climate shift of the Pacific Ocean. *Oceanography* 7:21-26.

Mohn, R.K. 1993. Bootstrap estimates of ADAPT parameters, their projection in risk analysis and their retrospective patterns. Pp. 173-184 in S.J. Smith, J.J. Hunt, and D. Rivard (eds.) *Risk Evaluation and Biological Reference Points for Fisheries Management.* Canadian Special Publication of Fisheries and Aquatic Sciences 120.

Mohn, R.K., and R. Cook. 1993. *Workbook: Introduction to Sequential Population Analysis.* Scientific Council Studies 17. Northwest Atlantic Fisheries Organization, Dartmouth, Nova Scotia.

Mohn, R.K., G. Robert, and D.L. Roddick. 1987. Research sampling and survey design of Georges Bank scallops (*Placopecten magellanicus*). *J. Northwest Atl. Fish. Sci.* 8:117-122.

Moore, R.E. 1966. *Interval Analysis.* Prentice-Hall, Englewood Cliffs, N.J.

Moore, R.E. 1978. Methods and applications of interval analysis. *SIAM Studies on Applied Mathematics*, Vol. 2. Philadelphia, Pa.

Murawski, S.A., and J.T. Finn. 1988. Biological bases for mixed-species fisheries: Species co-distribution in relation to environmental and biotic variables. *Can. J. Fish. Aquat. Sci.* 45:1720-1734.

Murawski, S., K. Mays, and D. Christensen. 1994. Fishery Observer Program. Pp. 35-41 in *NEFSC Status of the Fishery Resources off the Northeastern United States for 1994.* NOAA Technical Memorandum NMFS-NE-108. National Marine Fisheries Service, Woods Hole, Mass.

Myers, R.A., and N.G. Cadigan. 1995. Statistical analysis of catch-at-age data with correlated errors. *Can. J. Fish. Aquat. Sci.* 52:1265-1273.

Myers, R.A., A.A. Rosenberg, P.M. Mace, N. Barrowman, and V.R. Restrepo. 1994. In search of thresholds for recruitment overfishing. *ICES J. Mar. Sci.* 51:191-205.

Myers, R.A., N.J. Barrowman, J.A. Hutchings, and A.A. Rosenberg. 1995. Population dynamics of exploited fish stocks at low population levels. *Science* 269:1106-1108.

Nandram, B., J. Sedransk, and S.J. Smith. 1995. *Order Restricted Bayesian Estimation of the Age Composition of a Fish Population.* Technical report. Worcester Polytechnic Institute, Department of Mathematical Sciences, Worcester, Mass.

Nandram, B., J. Sedransk, and S.J. Smith. 1997. Order restricted Bayesian estimation of the age composition of a population of Atlantic cod. *J. Amer. Stat. Assoc.* 92:33-40.

National Marine Fisheries Service (NMFS). 1995. *Fisheries Statistics of the United States, 1994.* Current Fishery Statistics No. 9400. U.S. Department of Commerce, Washington, D.C.

National Marine Fisheries Service (NMFS). 1996. *Our Living Oceans. Report of the Status of U.S. Living Marine Resources, 1995.* NOAA Technical Memorandum. NMFS-F/SPO-19. U.S. Department of Commerce, Washington, D.C.

National Oceanic and Atmospheric Administration (NOAA). 1990. *50 Years of Population Change Along the Nation's Coasts: 1960-2010.* U.S Department of Commerce, Washington, D.C.

National Research Council (NRC). 1992. *Oceanography in the Next Decade: Building New Partnerships.* National Academy Press, Washington, D.C.

National Research Council (NRC). 1994a. *Improving the Management of U.S. Marine Fisheries.* National Academy Press, Washington, D.C.

National Research Council (NRC). 1994b. *An Assessment of Atlantic Bluefin Tuna.* National Academy Press, Washington, D.C.

National Research Council (NRC). 1996. *Bering Sea Ecosystem.* National Academy Press, Washington, D.C.

Nielsen, L.A. 1983. Variation in the catchability of yellow perch in an otter trawl. *Trans. Amer. Fish. Soc.* 112:53-59.

Ona, E. 1990. Physiological factors causing natural variations in acoustic target strength of fish. *J. Mar. Biol. Ass. U.K.* 70:107-127.

Ona, E., and O.R. Godø. 1990. Fish reaction to trawling noise: The significance for trawl sampling. *Rapp. P.-V. Cons. Int. Explor. Mer.* 189:159-166.

Ona, E., and R.B. Mitson. 1996. Acoustic sampling and signal processing near the seabed: The deadzone revisited. *ICES J. Mar. Sci.* 53:677-690.

Orensanz, J.M., J. Armstrong, D. Armstrong, and R. Hilborn. In press. Decline of the crustacean fisheries from the greater Gulf of Alaska: An historical perspective. *Rev. Fish Biol. Fisheries.*

Parma, A.M. 1993. Retrospective catch-at-age analysis of Pacific halibut: Implications on assessment of harvesting policies. Pp. 247-265 in G. Kruse, D.M. Eggers, R.J. Marasco, C. Pautzke, and T.J. Quinn II (eds.), *Proceedings of the International Symposium on Management Strategies for Exploited Fish Populations.* Alaska Sea Grant College Program Report No. 93-02. University of Alaska, Fairbanks.

Parma, A.M., and P.J. Sullivan. 1996. Changes to stock assessment methodology. *Report of Assessment and Research Activities, International Pacific Halibut Commission 1995.* International Pacific Halibut Commission, Seattle, Wash.

Pauly, D. 1980. On the interrelationships between natural mortality, growth parameters, and mean environmental temperature in 175 fish stocks. *J. Cons. Int. Explor. Mer.* 39(2):175-192.

Pawson, M.G. 1991. The relationship between catch, effort and stock size in put-and-take trout fisheries, its variability and application to management. Pp. 72-80 in Cowx (ed.), *Catch-Effort Sampling Strategies.* Fishing News Books, Oxford, UK.

Pearse, P.H., and C.J. Walters. 1992. Harvesting regulation under quota management systems for ocean fisheries. *Mar. Pol.* 16:167-182.

Pelletier, D., and P. Gros. 1991. Assessing the impact of sampling errors on model-based management advice: Comparison of equilibrium yield per recruit variance estimators. *Can. J. Fish. Aquat. Sci.* 48: 2129-2139.

Pelletier, D., and A.M. Parma. 1994. Spatial distribution of Pacific Halibut (*Hippoglossus stenolepis*): An application of geostatistics to longline survey data. *Can. J. Fish. Aquat. Sci.* 51:1506-1518.

Perry, R.I., and S.J. Smith. 1994. Identifying habitat associations of marine fishes using survey data: An application to the Northwest Atlantic. *Can. J. Fish. Aquat. Sci.* 51:589-602.

Peterman, R.M., and G.J. Steer. 1981. Relation between sport-fishing catchability coefficients and salmon abundance. *Trans. Amer. Fish. Soc.* 110:585-593.

Pickett, G.D., and M.G. Pawson. 1991. A logbook scheme for monitoring fish stocks: An example from the UK bass (*Dicentrarchus labrax* L.) fishery. In Cowx (ed.), *Catch-Effort Sampling Strategies.* Fishing News Books, Oxford, UK.

Pimm, S.L. 1984. *Food Webs.* Chapman and Hall, London.

Polovina, J.J., G.T. Mitchum, N.E. Graham, M.P. Craig, E.E. DeMartini, and E.N. Flint. 1994. Physical and biological consequences of a climate event in the central North Pacific. *Fish. Oceanogr.* 3:15-21.

Pollock, K.H., C.M. Jones, and T.L. Brown. 1994. *Angler Survey Methods and Their Application in Fisheries Management.* American Fisheries Society Special Publication 25, Bethesda, Md.

Prager, M.H. 1994. A suite of extensions to a nonequilibrium surplus-production model. *Fish. Bull.* 92:374-389.

Punt, A.E., and D.S. Butterworth. 1993. Variance estimates for fisheries assessments: Their importance and how best to evaluate them. Pp. 145-162 in S.J. Smith, J.J. Hunt, and D. Rivard (eds.), *Risk Evaluation and Biological Reference Points for Fisheries Management.* Canadian Special Publication of Fisheries and Aquatic Sciences 120.

Punt, A.E., and R. Hilborn. 1997. Fisheries stock assessment and decision analysis: The Bayesian approach. *Rev. Fish Biol. Fish.* 7:1-29.

Quinn, T.J., and J.S. Collie. 1990. Alternative population models for eastern Bering Sea pollock. Pp. 243-257 in Loh-Lee Low (ed.). Proceedings of the Symposium on Application of Stock Assessment Techniques to Gadids. *Int. N. Pac. Fish. Comm. Bull.* No. 50.

Quinn, T.J., II. 1985. Management of the North Pacific Halibut Fishery. I. Sampling Considerations in Real-time Fishery Management. Washington Sea Grant Technical Report WSG-85-1:36-48.

Quinn, T.J., II, and R.B. Deriso. In press. *Quantitative Fish Dynamics*. Oxford University Press, New York.

Quinn, T.J., II, and N.J. Szarzi. 1993. Determination of sustained yield in Alaska's recreational fisheries. Pp. 61-84 in G. Kruse, D.M. Eggers, R.J. Marasco, C. Pautzke, and T.J. Quinn II (eds.), *Proceedings of the International Symposium on Management Strategies for Exploited Fish Populations*. Alaska Sea Grant College Program Report No. 93-02. University of Alaska, Fairbanks.

Quinn T.J., II, E.A. Best, L. Bijsterveld, and I.R. McGregor. 1983. Port sampling for age composition of Pacific halibut landings. Pp. 194-205 in W.G. Doubleday and D. Rivard (eds.), *Sampling Commercial Catches of Marine Fish and Invertebrates*. Canadian Special Publication of Fisheries and Aquatic Sciences 66.

Quinn, T.J., II, R. Fagen, and J. Zheng. 1990. Threshold management policies for exploited populations. *Can. J. Fish. Aquat. Sci.* 47:2016-2029.

Raftery, A.E., G.H. Givens, and J.H. Zeh. 1995. Inference from a deterministic population dynamics model for bowhead whales. *J. Am. Statist. Assoc.* 90:402-416.

Reed, W.J. 1974. A stochastic model for the economic management of a renewable animal resource. *Math. Biosc.* 22:313-337

Restrepo, V.R., and J.E. Powers. 1991. A comparison of three methods for handling the "plus" group in virtual population analysis in the presence of ageing errors. *International Commission for the Conservation of Atlantic Tunas Collective Volume of Scientific Papers* 35:346-354.

Restrepo, V.R., J.M. Hoenig, J.E. Powers, J.W. Baird, and S.C. Turner. 1992. A simple simulation approach to risk and cost analysis, with applications to swordfish and cod fisheries. *Fish. Bull.* 90:736-748.

Richards, L.J., and J.T. Schnute. 1992. Statistical models for estimating CPUE from catch and effort data. *Can. J. Fish. Aquatic Sci.* 49:1315-1327.

Ricker, W.E. 1975. Computation and interpretation of biological statistics of fish populations. *Bull. Fish. Res. Bd. Can.* 191.

Robert, G., G.A.P. Black, and M.A.E. Butler. 1994. *Georges Bank Scallop Stock Assessment—1994*. DFO Atlantic Fisheries Research Document 94/97, Canadian Department of Fisheries and Oceans, Halifax, Nova Scotia.

Robson, D.S. 1991. The roving creel survey. *Amer. Fish. Soc. Symp.* 12:19-24.

Robson, D.S., and C. Jones. 1989. The theoretical basis of an access site angler survey design. *Biometrics* 45:83-98.

Rose, G.A., and W.C. Legget. 1991. Effects of biomass-range interactions on catchability of migratory demersal fish by mobile fisheries: An example of Atlantic cod (*Gadus morhua*). *Can. J. Fish. Aquat. Sci.* 48:843-848.

Rosenberg, A., P. Mace, G. Thompson, G. Darcy, W. Clark, J. Collie, W. Gabriel, A. MacCall, R. Methot, J. Powers, U. Restrepo, T. Wainwright, L. Botsford, J. Hoenig, and K. Stakes. 1994. *Scientific Review of Definitions of Overfishing in U.S. Fishery Management Plans*. NOAA Technical Memorandum NMFS-F/SPO-17. National Oceanic and Atmospheric Administration, Washington, D.C.

Rosenberg, A.A., and V.R. Restrepo. 1994. Uncertainty and risk evaluation in stock assessment advice for U.S. marine fisheries. *Can. J. Fish. Aquat. Sci.* 51:2715-2720.

Roughgarden, J. 1975. A simple model for population dynamics in stochastic environments. *Amer. Natur.* 109:713-736.

Royer, T.C. 1993. High-latitude oceanic variability associated with the 18.6-year nodal tide. *J. Geophys. Res.* 98:4639-4644.

Russell, F.S. 1973. A summary of the observations on the occurrence of the planktonic stages of fish off Plymouth 1924-72. *J. Mar. Biol. Assoc. U.K.* 53:347-355.

Saila, S.B. 1992. Application of fuzzy graph theory to successional analysis of a multispecies trawl fishery. *Trans. Amer. Fish. Soc.* 121(2):211-233.

Saila, S.B. 1993. The use of multivariate trend analysis to provide preliminary multispecies management advice. Pp. 493-506 in G. Kruse, D.M. Eggers, R.J. Marasco, C. Pautzke, and T.J. Quinn II (eds.), *Proceedings of the International Symposium of Management Strategies for Exploited Fish Populations*. Alaska Sea Grant College Program Report No. 93-02. University of Alaska, Fairbanks.

Saila, S.B., E. Lorda, J.D. Miller, R.A. Sher, and W.H. Howell. In press. Equivalent adult estimates for egg, larval, and juvenile fish losses at Seabrook Station using fuzzy logic to represent parametric uncertainty. *N. Am. J. Fish. Manage.*

Sakuramoto, K. 1995. A method to estimate related recruitment from catch-at-age data using fuzzy control theory. *Fish. Sci.* 61:401-405.

Schaefer, M.B. 1954. Some aspects of dynamics of populations important to the management of commercial marine fisheries. *Int. Am. Trop. Tuna Comm. Bull.* 1:25-56.

Schnute, J. 1985. A general theory for the analysis of catch and effort data. *Can. J. Fish. Aquat. Sci.* 42: 414-429.

Schnute, J. 1987. A general fishery model for a size-structured fish population. *Can. J. Fish. Aquat. Sci.* 44: 924-940.

Sen, A.R. 1985. Methodological problems in sampling commercial rockfish landings. *Fish. Bull.* 84:409-421.

Serchuk, F.M., and S.E. Wigley. 1986. Evaluation of USA and Canadian research vessel surveys for sea scallops (*Placopecten magellanicus*) on Georges Bank. *J. Northwest Atl. Fish. Sci.* 7:1-13.

Serns, S.L. 1986. Cohort analysis as an indication of walleye year-class strength in Escanaba Lake, Wisconsin, 1956-1974. *Trans. Amer. Fish. Soc.* 115:849-852.

Shardlow, T.F. 1993. Components analysis of a density-dependent catchability coefficient in a salmon hook and line fishery. *Can. J. Fish. Aquat. Sci.* 50:513-520.

Shepherd, J.G. 1982. A versatile new stock-recruitment relationship for fisheries and the construction of sustainable yield curves. *J. Cons. Int. Explor Mer* 40:67-75.

Shepherd, J.G. (ed.). 1991. *Special Session in Management Under Uncertainties*. North Atlantic Fisheries Organization Science Council Studies 16.

Shepherd, J.G., and J.W. Horwood. 1979. The sensitivity of exploited populations to environmental "noise," and the implications for management. *J. Cons. Int. Explor. Mer.* 38:318-323.

Showell, M.A., and M.C. Bourbonnais. 1994. Status of the Scotian shelf silver hake populations in 1993 with projections to 1995. Northwest Atlantic Fisheries Organization SCR Doc. 94/32.

Simard, Y., P. Legendre, G. Lavoie, and D. Marcotte. 1992. Mapping, estimating biomass and optimizing sampling programs for spatially autocorrelated data: Case study of the northern shrimp (*Pandalus borealis*). *Can. J. Fish. Aquat. Sci.* 48:32-45.

Simmonds, E.J., and R.J. Fryer. 1996. Which are better, random or systematic acoustic surveys? A simulation using North Sea herring as an example. *ICES J. Mar. Sci.* 53:39-50.

Simmonds, E.J., N.J. Williamson, F. Gerlotto, and A. Aglen. 1992. Acoustic survey design and analysis procedure: A comprehensive review of current practice. ICES Cooperative Research Report 187.

Sinclair, A. 1992. Fish distribution and partial recruitment: The case of eastern Scotian Shelf cod. *J. Northwest Atl. Fish. Sci.* 13:15-24.

Sinclair, A., D. Gascon, R. O'Boyle, D. Rivard, and S. Gavaris. 1991. Consistency of some Northwest Atlantic groundfish stock assessments. *NAFO Scientific Council* 16:59-77.

Sissenwine, M.P., and J.G. Shepherd. 1987. An alternative perspective on recruitment overfishing and biological reference points. *Can. J. Fish. Aquat. Sci.* 44:913-918.

Smale, M.J., and A.E. Punt. 1991. Age and growth of the red steenbras *Petrus rupestris* (Pices Sparidae) on the southeast coast of South Africa. *S. Afr. J. Mar. Sci.* 9:249-259.

Smith, P.J. 1989. Is two-phase sampling really better for estimating age composition? *J. Amer. Stat. Assoc.* 84:916-921.

Smith, P.J., and J. Sedransk. 1982. Bayesian optimization of the estimation of the age composition of a fish population. *J. Amer. Stat. Assoc.* 77:707-713.

Smith, S.J., and S. Gavaris. 1993a. Improving the precision of fish abundance estimates of eastern Scotian Shelf cod from bottom trawl surveys. *N. Am. J. Fish. Manage.* 13:35-47.

Smith, S.J., and S. Gavaris. 1993b. Evaluating the accuracy of projected catch estimates from sequential population analysis and trawl survey abundance estimates. Pp. 163-172 in S.J. Smith, J.J. Hunt, and D. Rivard (eds.), *Risk Evaluation and Biological Reference Points for Fisheries Management*. Canadian Special Publication of Fisheries and Aquatic Sciences 120.

Smith, S.J., J.J. Hunt, and D. Rivard (eds.). 1993. Risk evaluation and biological reference points for fisheries management. Canadian Special Publication of Fisheries and Aquatic Sciences 120.

Smith, S.J., and F.H. Page. 1996. Associations between Atlantic cod (*Gadus morhua*) and hydrographic variables: Implications for the management of the 4VsW cod stock. *ICES J. Mar. Sci.* 53:597-614.

Smith, S.J., and G. Robert. 1997. Getting more out of your survey information: An application to Georges Bank scallops (*Placopecten magellanicus*). In G. Jamieson (ed.), *North Pacific Symposium on Invertebrate Stock Assessment and Management*, Canadian Special Publication of Fisheries and Aquatic Sciences 125.

Smith, S.J., R.I. Perry, and L.P. Fanning. 1991. Relationships between water mass characteristics and estimates of fish population abundance from trawl surveys. *E.M.A.* 17:227-245.

Soutar, A., and J.D. Isaacs. 1974. Abundance of pelagic fish during the 19th and 20th centuries as recorded in anaerobic sediments off California. *Fish. Bull.* 72:257-273.

Southward, A.J., S.J. Hawkins, and M.T. Burrows. 1995. Seventy years' observations of changes in distribution and abundance of zooplankton and intertidal organisms in the western English Channel in relation to rising sea temperature. *J. Therm. Biol.* 20:127-155.

Spencer, P.D. 1997. Optimal harvesting of fish populations with nonlinear rates of predation and autocorrelated environmental variability. *Can. J. Fish. Aquat. Sci.* 54:59-74.

Stanley, R.D. 1992. Bootstrap calculation of catch-per-unit-effort variance from trawl logbooks: Do fisheries generate enough observations for stock assessments? *N. Am. J. Fish. Manage.* 12:19-27.

Starr, R.M., D.S. Fox, M.A. Hixon, B.N. Tissot, G.E. Power, and W.H. Barss. 1995. Comparison of submersible-survey and hydroacoustic-survey estimates of fish density on a rocky bank. *Fish. Bull.* 94:113-123.

Stevenson, S.C. 1983. A review of sampling commercial groundfish catches in Newfoundland. Pp. 29-38 in W.G. Doubleday and D. Rivard (eds.), *Sampling Commercial Catches of Marine Fish and Invertebrates*. Canadian Special Publication of Fisheries and Aquatic Sciences 66.

Sullivan, P.J. 1991. Stock abundance estimation using depth-dependent trends and spatially correlated variation. *Can. J. Fish. Aquat. Sci.* 48:1691-1703.

Sullivan, P.J. 1992. A Kalman filter approach to catch-at-length analysis. *Biometrics* 48:237-257.

Summerfelt, R.C., and G.E. Hall (eds.). 1987. *Age and Growth of Fish*. Iowa State University Press, Ames.

Swain, D.P., and D.L. Kramer. 1995. Annual variation in temperature selection by Atlantic cod *Gadus morhua* in the southern Gulf of St. Lawrence, Canada and its relation to population size. *Mar. Ecol. Prog. Ser.* 116:11-23.

Swain, D.P., and A.F. Sinclair. 1994. Fish distribution and catchability: What is the appropriate measure of distribution? *Can. J. Fish. Aquat. Sci.* 51:1054-56.

Szarzi, N.J., T.J. Quinn II, and D.N. McBride. 1995. Assessment of shallow-water clam resources: Case study of razor clams, eastern Cook Inlet, Alaska. *ICES Mar. Sci. Symp.* 199:274-286.

Thompson, G.G. 1994. Confounding of gear selectivity and the natural mortality rate in cases where the former is a nonmonotone function of age. *Can. J. Fish. Aquat. Sci.* 51:2654-2664.

Thompson, S.K. 1992. *Sampling*. John Wiley & Sons, New York.

Thompson, S.K., and G.A.F. Seber. 1996. *Adaptive Sampling*. John Wiley & Sons, N.Y.

Traynor, J.J., and N.J. Williamson. 1983. Target strength measurements of walleye pollock (*Theragra chalcogramma*) and a simulation study of the dual beam method. *FAO Fisheries Report* 300:112-124.

Traynor, J.J., W.A. Karp, M. Furusawa, T. Sasaki, K. Teshima, T.M. Sample, N.J. Williamson, and T. Yoshimura. 1990. Methodology and biological results from surveys of walleye pollock (*Theragra chalcogramma*) in the eastern Bering Sea and Aleutian Basin in 1988. Pp. 69-99 in L.L. Low (ed.), *Proceedings of the Symposium on Application of Stock Assessment Techniques to Gadids*. International North Pacific Fisheries Commission, Bulletin Number 50.

United Nations. 1995. Agreement for the Implementation of the Provisions of the United Nations Convention on the Law of the Sea of 10 December 1982 Relating to the Conservation and Management of Straddling Fish Stocks and Highly Migratory Fish Stocks. United Nations General Assembly, New York.

U.S. GLOBEC. 1995. *Global Ocean Ecosystems Dynamics and Climate Change, A Long Range Science Plan, 1995-2005*. Report Number 12. April 1995, University of California, Berkeley.

Venrick, E.I., J.A. McGowan, D.R. Cayan, and T.L. Hayward. 1987. Climate and chlorophyll a: Long-term trends in the central North Pacific Ocean. *Science* 238:70-72.

Vetter, E.F. 1988. Estimation of natural mortality in fish stocks: A review. *U.S. Fish. Bull.* 86:25-43.

Wade, D.L., C.M. Jones, D.S. Robson, and K. Pollock. 1991. Computer simulation techniques to assess bias in the roving-creel-survey estimator. *Amer. Fish. Soc. Symp.* 12:40-46.

Waksberg, J. 1978. Sampling methods for random digit dialing. *J. Amer. Stat. Assoc.* 73:40-46.

Walsh, S.J. 1991. Diel variation in availability and vulnerability of fish to a survey trawl. *J. Appl. Ichthyology* 7:147-159.

Walters, C.J. 1986. *Adaptive Management of Renewable Resources*. Macmillan, New York.

Walters, C.J. 1996. Computers and the future of fisheries. Pp. 223-238 in B.A. Megrey and E. Moksness (eds.), *Computers in Fisheries Research*. Chapman and Hall, London.

Walters, C.J., and J.S. Collie. 1989. An experimental strategy for groundfish management in the face of large uncertainty about stock size and production. Pp. 13-25 in Canadian Special Publication of Fisheries and Aquatic Sciences 108.

Walters, C.J., and D. Ludwig. 1987. Adaptive management of harvest rates in the presence of a risk averse utility function. *Nat. Res. Model.* 1:321-337.

Walters, C.J., and A.M. Parma. 1996. Fixed exploitation rate strategies for coping with effects of climate change. *Can. J. Fish. Aquat. Sci.* 53:148-158.

Walters, C., and P.H. Pearse. 1996. Stock information requirements for quota management systems in commercial fisheries. *Rev. Fish Biol. Fisheries* 6:21-42.

Weithman, A.S. 1991. Telephone survey preferred in collecting data statewide. *Amer. Fish. Soc. Symp.* 12:271-280.

Winters, G.H., and J.P. Wheeler. 1985. Interaction between stock area, stock abundance, and catchability coefficient. *Can. J. Fish. Aquat. Sci.* 42:989-998.

Wooster, W.S., and A.B. Hollowed. 1991. Decadal scale changes in the Northeast Pacific Ocean. *Northwest Environmental J.* 7:361-363.

Zadeh, L. 1978. Fuzzy sets as a basis for a theory of possibility. *Fuzzy Sets and System* 1:3-28.

Zheng, J., M.C. Murphy, and G.H. Kruse. 1995a. A length-based population model and stock-recruitment relationships for red king crab, *Paralithodes camtschaticus*, in Bristol Bay, Alaska. *Can. J. Fish. Aquat. Sci.* 52:1229-1246.

Zheng, J., M.C. Murphy, and G.H. Kruse. 1995b. Updated length-based population model and stock-recruitment relationships for red king crab in Bristol Bay, Alaska. *Alaska Fish. Res. Bull.* 2(2):114-124.

Zheng, J., M.C. Murphy, and G.H. Kruse. 1996. A catch-length analysis for crab populations. *Fish. Bull.* 94:576-588.

Zwanenburg, K.C.T., and S.J. Smith. 1983. Comparison of finfish length-frequency distributions estimated from samples taken at sea and in port. Pp. 189-193 in W.G. Doubleday and D. Rivard (eds.), *Sampling Commercial Catches of Marine Fish and Invertebrates*. Canadian Special Publication of Fisheries and Aquatic Sciences 66.

Appendixes

A

Letter of Request

UNITED STATES DEPARTMENT OF COMMERCE
The Assistant Secretary for
Oceans and Atmosphere
Washington, D.C. 20230

FEB 2 4 1995

Dr. William J. Merrell
Chairman, Ocean Studies Board
National Research Council
2101 Constitution Avenue, N.W.
Washington, D.C. 20418

RECEIVED

FEB 2 1995

Dear Dr. Merrell:

I am writing to confirm our recent discussions that the
Ocean Studies Board (OSB) will conduct a review of stock
assessment methods used as the scientific basis for fisheries
management.

As the demand for seafood and recreational opportunities
increase, more and more fishery resources become depleted, and
overcapitalization becomes more severe, it is not surprising that
stock assessments are being subjected to unprecedented scrutiny.
All those with an interest in fisheries management expect
management to be based on sound assessments. The National
Oceanic and Atmospheric Administration (NOAA) takes this
responsibility very seriously, and we believe the National
Research Council (NRC) can help us achieve it. Therefore, I am
requesting that the NRC undertake a review of the state-of-the-
art of stock assessments. Our goal is an authoritative report
that documents strengths and limitations of stock assessment
methods relative to the diverse data types and fisheries
management systems.

Since NOAA's fishery laboratories contain a large portion
of the country's most experienced stock assessment scientists,
we feel that it is important that the study draw on their
expertise. Therefore, I have asked Nancy Foster, Deputy
Assistant Administrator for Fisheries, to work with OSB staff
to formulate an appropriate strategy for a stock assessment
study.

Sincerely,

Douglas K. Hall

Douglas K. Hall

THE DEPUTY ADMINISTRATOR

B

Committee Biographies

Richard Deriso earned his Ph.D. in biomathematics from the University of Washington in 1978. He is cochair of the Committee on Fish Stock Assessment Methods. Dr. Deriso is the chief scientist of the Tuna-Billfish Program of the Inter-American Tropical Tuna Commission. He also serves as adjunct associate professor at the Scripps Institution of Oceanography and affiliate associate professor of fisheries at the University of Washington. Dr. Deriso served on the Ocean Studies Board (OSB) Committee to Review Atlantic Bluefin Tuna. His major research interests are in the areas of fisheries population dynamics, quantitative ecology, stock assessment, applied mathematics, and statistics.

Terrance Quinn earned a Ph.D. in biomathematics from the University of Washington in 1980. He is cochair of the Committee on Fish Stock Assessment Methods. Dr. Quinn has been an associate professor at the University of Alaska since 1985. He is a member of the Scientific and Statistical Committee of the North Pacific Fisheries Management Council and has recently served as consultant to the National Marine Fisheries Service (NMFS) and the Makah Indian Tribe in Neah Bay, Washington. Dr. Quinn also served on the OSB Committee on Fisheries and Committee to Review Atlantic Bluefin Tuna and was appointed to the OSB in 1995. His research interests are in the areas of fish population dynamics and management, applied statistics, and biometrics.

Jeremy Collie earned his Ph.D. in biological oceanography from the Massachusetts Institute of Technology-Woods Hole Oceanographic Institution Joint Program in Oceanography in 1985. Dr. Collie serves as associate professor of oceanography at the Graduate School of Oceanography, University of Rhode Island. His research interests include quantitative ecology with emphasis on population dynamics and production of marine animals, fish population dynamics and management, fish feeding and prey selection, predator-prey interactions, and recruitment of marine fish.

Ray Hilborn earned his Ph.D. in zoology from the University of British Columbia in 1974. Dr. Hilborn is professor at the School of Fisheries of the University of Washington. His main areas of research are resource management, population dynamics, systems analysis, and fisheries.

Cynthia Jones earned her Ph.D. in oceanography from the University of Rhode Island in 1984. She is associate professor in the Department of Biology at Old Dominion University. Dr. Jones' main areas of research are fisheries and population ecology.

Bruce Lindsay earned his Ph.D. in biomathematics from the University of Washington in 1978. He is professor of statistics at Pennsylvania State University. His research focuses on statistical methods in semiparametric models, with emphasis on maximum likelihood and minimum distance methods in mixture models and computation.

Ana Parma earned her Ph.D. in fisheries from the University of Washington in 1988. She is currently a population dynamicist for the International Pacific Halibut Commission. Her research focuses on population dynamics and adaptive fisheries management.

Saul Saila earned a Ph.D. in fishery biology from Cornell University in 1952. Dr. Saila recently retired from his position as professor of oceanography at the University of Rhode Island, where he had been employed since 1956. Dr. Saila served on the OSB Committee to Review Atlantic Bluefin Tuna. His research focused on the area of fish population dynamics.

Lynda Shapiro earned her Ph.D. from Duke University. She is professor of biology at the University of Oregon and director of the university's Institute of Marine Biology. She is a commissioner of the South Slough National Estuarine Research Reserve, serves on the executive committee of the National Association of Marine Laboratories, and holds a number of other board and committee positions with various professional and academic organizations. Dr. Shapiro served on the OSB Committee to Review Atlantic Bluefin Tuna, the Committee on the Arctic Research Vessel, and the Ocean Studies Board. Her research focuses on marine phytoplankton ecology.

Stephen J. Smith earned an M.Sc. degree in statistics from the University of Guelph, Canada, in 1979. He is a research scientist for the Department of Fisheries and Oceans of the Bedford Institute of Oceanography. Mr. Smith has served as chairperson of the Statistics, Sampling and Surveys Subcommittee of the Canadian Atlantic Fisheries Scientific Advisory Committee. Additionally, he has served on the editorial board of the *Canadian Journal of Fisheries and Aquatic Science* and on the board of directors of the Statistical Society of Canada as Atlantic Provinces representative, as well as being a member of a number of the society's committees. Currently, Mr. Smith is assistant editor of the *ICES Journal of Marine Science*. His primary research interests are in the field of resource management and modeling of marine fisheries, with a concentration in statistics.

Carl Walters earned a Ph.D. in fisheries from Colorado State University in 1969. Dr. Walters is professor of zoology and animal resource ecology at the University of British Columbia. His areas of research include the dynamics of ecological communities, application of mathematical models and computer simulation techniques to problems in resource ecology, and adaptive management of renewable resources.

C

Acronyms and Symbols

ADAPT	adaptive approach (age-structured)
ADF&G	Alaska Department of Fish and Game
ADMB	Autodifferentiation Model Builder
ASMFC	Atlantic States Marine Fisheries Commission
B	biomass
BRP	biological reference point
CLA	catch-length analysis
CPUE	catch per unit effort
CSA	catch-survey analysis
CV	coefficient of variation
EB	exploitable biomass
EEZ	exclusive economic zone
ENSO	El Niño-Southern Oscillation
F	instantaneous fishing mortality rate
FAO	Food and Agriculture Organization
GIS	geographic information system
GLOBEC	Global Ocean Ecosystems Dynamics program
GSMFC	Gulf States Marine Fisheries Commission
ICCAT	International Commission for the Conservation of Atlantic Tunas
ICES	International Council for the Exploration of the Sea
LBA	length-based analysis
LTPY	long-term potential yield

M	instantaneous natural mortality rate
MAP	maximum a posteriori
MRFSS	Marine Recreational Fisheries Statistics Survey
MSVPA	multispecies virtual population analysis
MSY	maximum sustainable yield
NMFS	National Marine Fisheries Service
NOAA	National Oceanic and Atmospheric Administration
NRC	National Research Council
OMB	Office of Management and Budget
OSB	Ocean Studies Board
PSMFC	Pacific States Marine Fisheries Commission
q	catchability
SAFE	Stock Assessment and Fishery Evaluation
SD	standard deviation
SPR	spawning biomass per recruit
S-R	stock-recruitment
SSB	spawning stock biomass
TAC	total allowable catch
VPA	virtual population analysis
YPR	yield per recruit
Z	total instantaneous mortality rate

D

Checklist for Stock Assessments

Table D.1 contains a checklist of items that should be included and/or considered in a stock assessment. The assumptions given in Table D.1 should be considered in choosing a model and specific parameter values.

TABLE D.1   Checklist for Conducting or Reviewing Stock Assessments

Step	Important Considerations

1.0 Stock Definition
    Stock structure
    Single or multispecies
	What is the spatial definition of a "stock"?
	Should the assessment be spatially structured or assumed to be spatially homogeneous?
	Choose single-species or multi-species assessment?
	Use tagging, micro-constituents, genetics, and/or morphometrics to define stock structure?

2.0 Data	

2.1 Removals
    Catch
    Discarding
    Fishing-induced mortality
	Are removals included in the assessment?
	Are biases and sampling design documented?

2.2 Indices of abundance	For all indices, consider whether an index is absolute or relative, sampling design, standardization, linearity between index and population abundance, what portion of stock is indexed (spawning stock, vulnerable biomass).
Catch per unit effort (CPUE)	What portions of the fleet should be included and how should data be standardized? How are zero catches treated? What assumptions are made about abundance in areas not fished? Spatial mapping of CPUE is especially informative.
Gear surveys (trawl, longline, pot)	Is gear saturation a problem? Does survey design cover the entire range of the stock? How is gear selectivity assessed?
Acoustic surveys	Validate species mix and target strength.
Egg surveys	Estimate egg mortality, towpath of nets, and fecundity of females.
Line transect, strip counting	

TABLE D.1   (Continued)

Step		Important Considerations
2.3	Age, size, and sex-structure information	
	Catch at age Weight at age Maturity at age Size at age Age-specific reproductive information	Consider sample design, sample size, high-grading selectivity, and ageing errors.
2.4	Tagging data	Consider both tag loss and shedding and tag return rates. Was population uniformly tagged or were samples recovered?
2.5	Environmental data	How should such data be used in the assessment? What are the dangers of searching databases for correlates?
2.6	Fishery information	Are people familiar with the fishery, who have spent time on fishing boats, consulted and involved in discussions of the value of different data sources?
3.0	Assessment Model	
3.1	Age-, size-, length-, or sex-structured model?	Are alternative structures considered?
3.2	Spatially explicit or not?	
3.3	Key model parameters	
	Natural mortality Vulnerability Fishing mortality Catchability	Are these parameters assumed to be constant or are they estimated? If they are estimated, are prior distributions assumed? Are they assumed to be time invariant?
	Recruitment	Is a relationship between spawning stock and recruitment assumed? If so, what variance is allowed? Is depensation considered as a possibility? Are environmentally driven reductions (or increases) in recruitment considered?
3.4	Statistical formulation	
	What process errors? What observation errors? What likelihood distributions?	If the model is in the form of weighted sum of squares, how are terms weighted? If the model is in the form of maximum likelihood, are variances estimated or assumed known?
3.5	Evaluation of uncertainty	
	Asymptotic estimates of variance Likelihood profile Bootstrapping Bayes posteriors	How is uncertainty in model parameters or between alternative models calculated? What is actually presented, a distribution or only confidence bounds?
3.6	Retrospective evaluation	Are retrospective patterns evaluated and presented?

TABLE D.1 (Continued)

Step	Important Considerations
4.0 Policy Evaluation	
4.1 Alternative hypotheses	What alternatives are considered: parameters for a single model or different structural models? How are the alternative hypotheses weighted? What assumptions are used regarding future recruitment, environmental changes, stochasticity, and other factors? Is the relationship between spawners and recruits considered? If so, do future projections include autocorrelation and depensation?
4.2 Alternative actions	What alternative harvest strategies are considered? What tactics are assumed to be used in implementation? How do future actions reflect potential changes in future population size? Is implementation error considered? Are errors autocorrelated? How does implementation error relate to uncertainty in the assessment model?
4.3 Performance indicators	What is the real "objective" of the fishery? What are the best indicators of performance? What is the time frame for biological, social, and economic indices? How is "risk" measured? Are standardized reference points appropriate? Has overfishing been defined formally?
5.0 Presentation of Results	How are uncertainties in parameters and model structure presented? Can decision tables be used to summarize uncertainty and consequences? Is there explicit consideration of the trade-off between different performance indicators? Do the decisionmakers have a good understanding of the real uncertainty in the assessment and the trade-offs involved in making a policy choice?

E

Description of Age-Structured Simulation Model

˙ Jie Zheng and Terry Quinn*

Each of the five data sets contains statistics from the fishery: reported catch, effort, and age composition. Survey data are summarized as a relative index along with survey age composition. Simulations were initiated in year -15 in a pristine population condition and ended in year 30. Fishery and survey data from years 1 to 30 were given to the analysts for stock assessment analyses. The population contains ages 1 to 15, where age 15 represents fish of 15 years age and older, although fish older than 15 are uncommon. Parameters are summarized in Tables E.1 and E.2.

In the description below, *a* denotes age and *t* denotes time. The age-structured model used to create simulated data sets has the following features:

1. Natural Mortality M_t: This is a random variable drawn as an annual value from a uniform distribution ranging from 0.18 to 0.27; the natural mortality is constant over age.
2. Growth: Mean weight at age follows an isometric von Bertalanffy curve

$$W_a = W_\infty \{1 - \exp[-\kappa(a - t_0)]\}^{\beta_1},$$

where $W_\infty = 5000$ g, $\kappa = 0.3$ per year, $\beta_1 = 3$, and $t_0 =$ year 0.

3. Maturity: The maturity relationship is a logistic-shaped function

$$m_a = 1/[1 + \exp(-b_m(a - a_m))].$$

The b_m parameter is constant over the five data sets and equal to 1.65 per year. The a_m parameter is the age at 50% maturity and varies among data sets. Both the growth and the maturity relationships are based on true age.

4. Spawner-Recruit Model: Spawning (*S*) is assumed to take place at the start of the year. Recruitment (*R*) is

---

*Jie Zheng and Terry Quinn constructed the Excel model using committee specifications and created the simulated data sets using the Excel model.

TABLE E.1 Summary of Simulated Population Parameters for the Five Data Sets

Parameter	Data Set 1	Data Set 2	Data Set 3	Data Set 4	Data Set 5
M_t	0.18-0.27	0.18-0.27	0.18-0.27	0.18-0.27	0.18-0.27
W_∞	5000	5000	5000	5000	5000
κ	0.3	0.3	0.3	0.3	0.3
β_1	3	3	3	3	3
t_0	0	0	0	0	0
a_m	7	8	9	8	7
b_m	1.65	1.65	1.65	1.65	1.65
α	3.8742	4.8276	6.1711	4.8276	3.8742
β	4.54×10^{-3}	4.53×10^{-3}	4.82×10^{-3}	3.23×10^{-3}	2.27×10^{-3}
σ	0.6	1.0	0.9	0.7	0.8
ρ	0.5	0.5	0.5	0.5	0.5
R_0	800	1000	1200	1400	1600
ε_{10}	−1.0	−1.0	−1.0	−1.0	1.0
a_f ($t = -15$ to 20)	5	5	5	5	5
b_f ($t = -15$ to 20)	0.8	0.8	0.8	0.8	0.8
a_f ($t = 21$ to 30)	5 to 3	5 to 3	5 to 3	5	5
b_f ($t = 21$ to 30)	1.0	1.0	1.0	0.8	0.8
σ_f	0.2	0.2	0.2	0.2	0.2
q_0	0.01	0.01	0.01	0.01	0.01
q_1	0.4	0.4	0.4	0.4	0.4
σ_q	0.2	0.2	0.2	0.2	0.2
r ($t = -15$ to 0)	0.0	0.0	0.0	0.0	0.0
r ($t = 1$ to 30)	0.0135	0.0135	0.0135	0.0135	0.0135
E_0	2738.2	2849.4	2922.8	3486.8	4150.3
E_1 ($t = -14$ to 1)	0.5	0.2	0.4	0.4	1.7
E_1 ($t = 2$ to 30)	0.68	0.65	0.8	0.8	0.3
Underreporting	0%	30%	0%	0%	0%
α'	3.2322	3.2322	3.2322	3.2322	3.2322
β'	3.3382	3.3382	3.3382	3.3382	3.3382
q^*_0 ($t = 1$ to 15)	2.0×10^{-6}	2.0×10^{-6}	2.0×10^{-6}	2.0×10^{-6}	2.0×10^{-6}
q^*_0 ($t = 16$ to 30)	2.0×10^{-6}	2.0×10^{-6}	4.0×10^{-6}	2.0×10^{-6}	2.0×10^{-6}
σ^*	0.3	0.3	0.3	0.3	0.3

defined as abundance at age 1; thus, $R_t = N_{1,t}$. Recruitment is assumed to follow a Beverton-Holt model with autocorrelated errors:

$$R_{t+1} = [\alpha \, S_t/(1 + \beta \, S_t)] \exp[e_t + u_t - 0.5 \, \sigma^2],$$
$$e_t = \rho \, e_{t-1} + z, \, z \sim N(0, \sigma'^2), \, \rho = 0.5, \, \sigma' = \sigma \, (1 - \rho^2)^{0.5},$$
$$u_t = -0.5 \, \sigma^2 \, \Sigma \, \rho^{t+16} \text{ for } t = -15 \text{ to } 9, \, u_t = -0.5 \, \sigma^2 \, \Sigma \, \rho^{t-11} \text{ for } t = 12 \text{ to } 30,$$
$$u_{10} = 0.5 \, \sigma^2, \text{ and } u_{11} = 0.$$

The error ε_{10} was set to be negative (−1) for declining data sets 1-4 and to be positive (1) for data set 5 (recovering

TABLE E.2   Summary of Reference Fishing Mortality and Effort and Associated Exploitable Biomass and Yield for the Five Data Sets

Parameter	Data Set 1	Data Set 2	Data Set 3	Data Set 4	Data Set 5
$F_{40\%}$	0.131	0.119	0.109	0.158	0.182
$E_{40\%}$	980.14	817.08	914.46	1405.12	1598.84
$B_{40\%}$	2596.32	2269.25	3183.21	3466.97	3416.25
$Y_{40\%}$	289.22	229.61	295.45	464.35	521.31
F_{MSY}	0.196	0.158	0.151	0.252	0.280
E_{MSY}	1222.8	949.1	1096.37	1826.68	2024.99
B_{MSY}	1924.38	1819.5	2479.86	2476.98	2473.06
MSY	312.20	240.46	315.18	512.68	563.53
F_{MSY1}	0.297	0.225	0.208	0.252	0.280
E_{MSY1}	2302.01	1722.76	1939.33	2740.02	3037.49
B_{MSY1}	1410.40	1379.94	1911.31	2476.98	2473.06
$MSY1$	339.53	257.13	331.41	512.68	563.53

stock) to force recruitment to be low for data sets 1-4 and high for data set 5. The above adjustments were done so that the expected value of recruitment follows the deterministic Beverton-Holt curve. Note that α and u_t can be combined as $\alpha_t = \alpha \exp(u_t)$.

Parameters σ, α, β, and R_0 (pristine, initial recruitment in millions) vary among data sets.

5. Population Abundance and Mortality Equations:

Abundance $N_{a+1,t+1} = N_{a,t} \exp(-Z_{a,t})$
Total mortality $Z_{a,t} = M_t + F_{a,t}$
Fishing mortality $F_{a,t} = s_a q_t E_t$

where s_a is selectivity, q_t is catchability, and E_t is fishing effort for the commercial fishery, described further below.

6. Catch Equation: Fishing and natural mortality occur continuously throughout the year, so that catch (C) follows from the Baranov equation:

$$C_{a,t} = N_{a,t} (F_{a,t}/Z_{a,t})[1 - \exp(-Z_{a,t})].$$

The catch for age 1 was small and reported as none.

7. Yield and Biomass: Yield (Y, catch in weight) and biomass (B, population size in weight) are obtained by multiplying catch and abundance (in numbers of fish) by average weight:

$$Y_{a,t} = C_{a,t} W_{a,t} \text{ and } B_{a,t} = N_{a,t} W_{a,t}.$$

8. Fishery Selectivity: One fishing gear was assumed for the simulation and the selectivity (S) for the fleet is a logistic function

$$s_a = 1/\{1 + \exp[(-b_f(a-a_f)]\} \exp(\varepsilon_t - 0.5\sigma_f^2).$$
$$\varepsilon_t \sim N(0, \sigma_f^2), \quad \sigma_f = 0.2 \text{ for all data sets.}$$

For data sets 1-3,

b_f = 0.8 and a_f = 5 for t = –15 to 20, and
b_f = 1.0 and a_f = 5 – 2 {1-exp[–0.45 $(t–20)$]} for t = 21 to 30.

For data sets 4-5,

b_f = 0.8 and a_f = 5 for t = –15 to 30.

For all data sets, a_f is equal to its expected value plus a value randomly between -1 and 1.

9. Fishery Catchability: Catchability (q) was assumed to have a exponential time trend and an allometric dependence on biomass:

$$q_t = q_0\ B^{q_1-1} \exp\left[r\,(t-t_0) + \varepsilon_t - 0.5\,\sigma_q^2\right]$$

for all data sets, where q_0 = 0.01, q_1 = 0.4, and t_0 = 0. Parameter r = 0 and $B_t = B_\infty$ (pristine exploitable biomass) for t = –15 to 0, and r = 0.0135 and B_t = exploitable biomass in year t for t = 1 to 30, so that catchability in year 30 is 50% higher than in year 1 for the same level of exploitable biomass B_t. Error component $\varepsilon_t \sim N(0, \sigma_q^2)$, σ_q = 0.2.

10. Fishing Effort: $E_t = E_0\,E_1\,v$, where v ranges from 0.75 to 1.25 randomly for all data sets, E_0 is a constant fishing effort, and E_1 is a constant. Both E_0 and E_1 were adjusted for different data sets to get desirable abundance trends over time. Reported effort is the true effort for all data sets.

11. Reported Catch: Total reported catch equals (true catch) × (1 – underreporting %) × u, where u ranges from 0.97 to 1.03 randomly each year. Underreported total catch and total yield are 30% for data set 2 and 0% for all other data sets. Reported yield in biomass is determined from landing reports, not as the sum of catch-age times weight-age. Reported catch in numbers is also not affected by age composition.

12. Survey Abundance: The survey occurs during a short period of time at the beginning of each year immediately after spawning, after recruitment, and before any mortality occurs. We follow the convention that birthdate is assigned at the beginning of each year just prior to spawning. The survey abundance index is calculated as

$$I_{a,t} = q_t^*\, S_a^*\, N_{a,t}$$

13. Survey Selectivity: The selectivity for the survey gear is a gamma function

$$s_a^* = \alpha^{\alpha'-1}\,\exp(-a/\beta')/[\beta'^{\alpha'-1}\Gamma(\alpha')],$$

where α' = 3.2322 and β' = 3.3382, resulting in $s_3^* = s_{15}^*$ = 0.5 and $\overset{*}{s}_7$ = 1.0.

14. Survey Catchability: The survey catchability is a constant multiplied by a random variable:

$$q_t^* = q_0^*\,\exp(\varepsilon_t - 0.5\,\sigma^{*2}),\ (\varepsilon_t \sim N(0,\sigma^{*2}),\ \sigma^* = 0.3,$$

where q_0^* = 0.000002 for t = 1 to 30, except for data set 3 for which q_0^* = 0.000002 for t = 1 to 15 and q_0^* = 0.000004 for t = 16 to 30.

15. Sample Size for Age Composition: Simple random samples of age composition were taken from the catch (n = 500) and from the survey (n = 200).

16. Ageing Error: Ageing error is the difference between the age reader's estimated age of a fish and its true

age. We assumed that a single reader aged all fish. Ageing error was generated with 0 bias at age 1, which increases linearly to −1 at age 15. The variation in ageing error was ~$N(0, \tilde{\sigma}^2)$, with a linear increase in standard deviation from $\tilde{\sigma} = 0$ for age 1 and $\tilde{\sigma} = 2$ for age 15. This resulted in a misclassification matrix (the age error matrix was provided to the analysts) whose elements are the probabilities $p_{i,a}$ that an individual of true age i is estimated to be age a.

17. Reference Fishing Mortality and Effort: Under the expected fishery selectivity and catchability in year 30, full-recruitment fishing mortality ($F_{40\%}$) and effort ($E_{40\%}$) in boat-days that reduce spawning biomass per recruit to 40% of the unfished level as well as the associated expected exploitable biomass ($B_{40\%}$) and yield ($Y_{40\%}$) in thousand metric tons, are given in Table E.2. Under the deterministic condition, full-recruitment fishing mortality (F_{MSY}) and effort (E_{MSY}) in boat-days at *MSY* (maximum sustainable yield), as well as the associated expected exploitable biomass (B_{MSY}) and *MSY* in thousand metric tons are also given in Table E.2.

Full-recruitment fishing mortality (F_{MSY1}) and effort (E_{MSY1}) in boat-days at *MSY* and the associated expected exploitable biomass (B_{MSY1}) and *MSY* in thousand metric tons were determined using the expected fishery selectivity and catchability in year 1 and deterministic conditions.

F

Letter to Analysts

March 19, 1996

Dear [Analysts]:

We are ready to embark on the simulation study discussed at the OSB Stock Assessment Committee meeting in January. The following analysts have agreed to work on the project (names and locations in original letter are deleted here):

Analyst 5, Analyst 2, and Analyst 3 —will do two delay-difference models (measurement error, measurement and process error) and a stock synthesis model

Analysts 6 and 7—will do a stock synthesis model and a Bayesian stock synthesis model

Analyst 1—will do a nonequilibrium production model

Analyst 4 will do an ADAPT model. As discussed at the January meeting, Analyst 4 would prefer to do this in a workshop setting with other colleagues. . . We have no objection to this and hope that NMFS will provide the support to enable this. We request that these teams do their work independently of each other.

Attached please find an Excel spreadsheet that contains data sets from 5 age-structured populations. Each data set contains information on many population parameters such as growth and maturity. There are also statistics from the fishery: reported catch and effort and age composition. A survey was conducted and summarized as a relative index along with survey age composition. Simple random samples of age composition were taken from the catch (n=500) and from the survey (n=200). Ageing error is present and the ageing error matrix is given to you. There are some further details below.

We would like you to analyze each data set in three ways if you can:

[A]: using CPUE as the only measure of relative abundance;
[B]: using only survey information;
[C]: using both CPUE and survey information.

This will allow us to address the question of whether surveys are important. We can label the analyses as 1A, 1B, 1C, 2A,..., 5C. IT IS CRITICAL THAT THE ''A'' AND ''B'' ANALYSES BE DONE FIRST AND ARE INDEPENDENT OF THE ''C'' ANALYSES. (i.e. Do not revise the A and B analyses based on what you come up with in the C analyses.)

As time permits, we would also like to get retrospective analyses of each data set and analysis. Ideally, we would like to get 15 retrospectives per analysis (i.e. years 1-16, ... years 1-30). These analyses should be done independently (i.e., please do not use results from years 1-30 to initialize the parameters for the retrospective analyses). We realize this may be optimistic, but given your time constraints we would like to have at least 5 retrospective analyses for each for the May meeting and more if you can do it.

We would like the results summarized as follows: summarized estimates of model parameters and model structure, estimated exploitable, mature, and total biomasses over time, average fishing mortality and exploitation rate over time, estimated recruitment (youngest age used) if part of the model, and selectivity gives for the fishery and survey. You are welcome to estimate a TAC or ABC; however if you do so, please tell us what approach you will use. As a default, we recommend $F_{40\%}$ be calculated for comparison.

Other model features:

1. Data from the fishery occurs over 30 years, t = 1,...,30. Age 15 represents a plus group but fish older than 15 are uncommon.
2. Natural mortality is unknown, may not be constant, but is in line with species with similar longevity.
3. Growth: Mean weight at age follows an allometric von Bertalanffy curve $W(a) = Wm [1 - \exp(-k(a - t0))]^b$. The parameters are given to you.
4. The maturity relationship is: $m(a) = 1/[1 + a \exp(-b\,a)]$ with given parameters.
5. The generation of recruitment is unknown to you.
6. Ageing error was generated with 0 bias at age 1 which increases linearly to -1 at age 15. The variation in ageing error was $\sim N(0, s^2)$, with a linear increase from s = 0 for age 1 and s = 2 for age 15.
7. For Set 3, a different vessel was used in years t = 16 to 30, which may or may not have altered survey catchability.
8. Catch Equation: Fishing occurs continuously throughout the year.
9. Reported yield in biomass is determined from landing reports, not as the sum of catch-age times weight-age. Reported catch in numbers is also not affected by age composition.

The survey occurs during a short period of time at the beginning of each year immediately after spawning. We follow the convention that birth-date is assigned at the beginning of each year just prior to spawning.

The ageing error measure is a comparison of the age reader's estimated age of fish to known true age. The comparison results in a misclassification matrix (the age error matrix is provided to the analysts) whose elements are the probability p(i,a), that an individual of true age "i" is said to be estimated age "a".

Sincerely,

Terry Quinn and Rick Deriso
Co-chairs, NRC Committee on Fish Stock Assessment Methods

NOTE: The following analysts agreed to be experimental units to apply the indicated methods:
(names deleted)

G

Retrospective Values

TABLE G.1   Retrospective Summaries of Terminal Year Exploitable Biomass

Data Set	End Year	Stock Assessment Models SS-P3 (F)	SS-P3 (B)	SS-P3 (S)	SS-P7 (F)	ADMB4 (F)	ADMB4 (B)	NRC ADAPT (F)	NRC ADAPT (B)	NRC ADAPT (S)	True Value	Absolute Change in True Value
1	15	2764	1454	1113	2793	938	742	1680	1073	820	1154	
1	16	2549	1292	1165	2493	1297	787	1896	972	918	689	−465
1	17	2170	849	777	2055	1201	663	1666	810	962	591	−98
1	18	2090	835	775	2076	986	663	1387	754	709	414	−177
1	19	2027	871	784	1514	985	724	1421	677	623	519	105
1	20	886	876	459	1624	492	415	1068	514	470	730	211
1	21	329	549	418	534	315	307	1126	623	538	677	−53
1	22	333	490	408	384	326	274	1174	572	498	853	176
1	23	310	441	296	326	353	292	1425	728	654	764	−89
1	24	386	373	274	363	413	202	1223	545	478	548	−216
1	25	649	406	375	544	446	227	986	494	436	240	−308
1	26	661	307	285	261	542	195	1367	438	371	289	49
1	27	654	236	222	180	567	146	1433	382	326	261	−28
1	28	339	211	190	182	491	150	1142	484	380	402	141
1	29	276	224	215	177	192	133	1213	440	323	249	−153
1	30	418	252	241	182	172	120	1476	378	298	276	27
2	15	2061	1001	938	1820	1714	1061	1443	1155	1141	1466	
2	16	2417	901	955	2731	2350	973	14,537	1302	1285	978	−488
2	17	670	751	731	737	1742	818	1929	1158	1109	1029	51
2	18	869	802	802	1035	1603	781	1323	1073	1051	791	−238
2	19	652	745	791	647	1109	638	1150	741	735	824	33
2	20	616	866	670	831	1202	632	1104	512	492	1055	231
2	21	864	855	703	1214	1328	579	1929	868	827	1082	27
2	22	857	972	677	1324	1914	807	6691	1098	1096	1332	250
2	23	618	628	371	663	995	467	2957	1015	978	1124	−208
2	24	452	305	248	456	660	333	1900	822	731	766	−358
2	25	1011	408	438	1047	511	282	1757	573	468	400	−366
2	26	720	263	301	628	672	228	1267	481	388	287	−113
2	27	689	165	201	111	492	122	1048	293	239	208	−79
2	28	279	162	229	100	411	159	861	403	366	440	232
2	29	614	256	292	200	495	205	881	436	381	239	−201
2	30	463	222	259	170	373	177	1008	434	330	346	107
3	15	7622	7139	5898	7622	5533	4564	4715	4218	3935	5033	
3	16	6365	6151	6024	6346	4027	4064	3972	4093	4032	4603	−430
3	17	5669	5480	5360	5680	4398	4536	4074	3697	3684	4388	−215
3	18	4817	4636	4525	4823	5286	5005	4700	4216	4205	3145	−1243
3	19	4178	4090	3943	4300	4730	4862	4547	4549	4458	2813	−332
3	20	1327	4816	3548	4848	1864	4752	4121	4653	4637	3379	566
3	21	1192	4664	1602	5112	1821	2938	4955	5079	5098	3794	415
3	22	1961	5097	2933	5017	2892	3761	6261	7590	7660	3946	152
3	23	1618	5047	2505	4509	2648	3621	5593	5658	5820	3426	−520
3	24	2451	4510	1882	1878	2901	2626	4589	5276	5262	3430	4
3	25	4818	4426	3701	4385	3298	2806	3990	4375	4484	2444	−986
3	26	4805	3904	3079	4186	3720	2300	3778	3594	3566	2170	−274
3	27	3930	3345	3039	3212	3895	2191	3449	3118	3093	1492	−678
3	28	1925	1201	1127	558	4097	1962	3627	3144	3077	1744	252
3	29	3522	1960	1746	2743	3962	1405	3371	2431	2272	1032	−712
3	30	2871	1808	1835	2089	4014	1393	3289	2286	2032	903	−129

TABLE G.1 (Continued)

| Data Set | End Year | Stock Assessment Models | | | | | | | | | True Value | Absolute Change in True Value |
		SS-P3 (F)	SS-P3 (B)	SS-P3 (S)	SS-P7 (F)	ADMB4 (F)	ADMB4 (B)	NRC ADAPT (F)	NRC ADAPT (B)	NRC ADAPT (S)		
4	15	1957	1087	1099	1584	2483	1180	3402	1593	1444	1693	
4	16	1218	803	832	1376	1844	904	2992	1244	1121	1049	−644
4	17	1445	774	772	1139	1869	874		967	861	947	−102
4	18	1432	752	771	1474	1758	774	2491	630	720	699	−248
4	19	1419	891	893	1573	1816	907	2593	1063	910	594	−105
4	20	560	724	528	1182	755	536	1815	829	778	669	75
4	21	432	563	479	730	575	415	1474	650	568	463	−206
4	22	250	508	444	220	501	387	1467	727	666	408	−55
4	23	386	414	397	426	678	382	1809	739	584	660	252
4	24	342	436	357	310	462	316	1370	800	665	469	−191
4	25	399	482	497	348	397	340	1537	776	576	218	−251
4	26	511	522	539	420	383	363	2058	869	702	339	121
4	27	469	522	533	420	394	358	1943	787	737	290	−49
4	28	315	393	357	322	351	296	2216	840	670	355	65
4	29	127	142	142	134	138	118	1372	466	246	97	−258
4	30	106	121	119	109	138	107	1019	362	225	115	18
5	15	3415	4334	3792	3458	1305	1660	2565	3124	3193	6055	
5	16	4632	4866	4120	4738	2200	1995	2729	3519	3560	4851	−1204
5	17	5273	6412	5938	5390	2564	2616	3212	3906	3969	6251	1400
5	18	6428	7318	6819	7234	3642	3760	3488	3650	3697	7004	753
5	19	6626	7204	6711	6777	3269	3811	3936	4350	4412	6344	−660
5	20	7038	8570	7357	7739	3131	3719	3769	4366	4441	7239	895
5	21	5880	7343	6567	7142	3164	3401	3430	3724	3498	7195	−44
5	22	5478	6870	6412	6489	3056	3224	3269	3984	4056	7164	−31
5	23	5829	6690	6408	6140	3195	3214	3158	4012	4042	7708	544
5	24	4502	6450	5908	4533	2106	2943	2369	4051	4077	8175	467
5	25	5379	6282	6464	6016	2988	2913	2996	3575	3615	6524	−1651
5	26	7018	6353	6688	7154	3182	2979	3384	3774	3805	7390	866
5	27	7258	6593	6671	7224	3482	3067	3030	3463	3511	5634	−1756
5	28	6494	5652	5583	6776	3650	2708	3085	3292	3307	6864	1230
5	29	6506	5276	5077	6592	3728	2792	2754	3077	3085	5001	−1863
5	30	5397	4477	4107	6333	3613	2427	2756	3005	2973	5158	157

NOTE: ADMB = Autodifferentiation Model Builder: SS = Stock Synthesis; ADAPT = adaptive approach.

H

Information from Interstate Marine Fisheries Commissions and States

To aid the committee's deliberations, information was requested from the Atlantic, Gulf, and Pacific States Marine Fisheries Commissions. The committee specifically asked the commissions for information about what methods their states use to assess marine fish stocks in state waters (to compare with methods used by National Marine Fisheries Service [NMFS] and the regional fishery management councils). The commissions were also asked to relay their concerns about existing stock assessment methods. This information is shown below; they are the views of the commissions and their staff and do not necessarily coincide with the views of the committee. The response from the Atlantic States Marine Fisheries Commission was received first and was forwarded to the other commissions.

COMMISSION AND STATE CONCERNS AND COMMENTS REGARDING STOCK ASSESSMENTS

Atlantic States Marine Fisheries Commission (ASMFC)

1. The complexity of many stock assessments models leads to a lack of understanding among those who may not have a scientific background. More effort is needed to provide less technical explanations of stock assessment models and results.

2. Discrepancies in model results due to incomplete characterization of input parameters lead to less effective and timely fisheries management, and lack of public confidence in the scientific process and expertise.

3. Probability analysis should be included as a part of all stock assessments to provide an indication of the level of achieving management and rebuilding goals.

4. Deficiencies in fisheries-dependent data bases limit the effectiveness of many stock assessments; i.e., catch at age, discards.

5. Deficiencies in information concerning critical model parameters may limit stock assessments; i.e., fishing and natural mortality, bias in catch data, stock distribution, and life history parameters.

6. Harvest regulations may affect input data. For example, the effect of size limits may be to truncate length-frequency data thereby affecting the model results.

7. Biases in model results may occur due to equilibrium assumptions; i.e., yield-per-recruit modeling.

TABLE H.1   Stock Assessment Methods Used by the Atlantic States Marine Fisheries Commission

Species	Stock Assessment Method(s)
American eel	None
American lobster	Cohort analysis, egg production per recruit, Delury
Atlantic menhaden	VPA (basic and separable)
Atlantic sturgeon	Stock recruitment, egg recruitment
Black sea bass (with MAFMC)	VPA (ADAPT and ICA)
Bluefish (with MAFMC)	VPA (ADAPT, Cagean or ICA), tagging
Croaker	None
Northern shrimp	Abundance indices
Red drum (with SAFMC)	VPA (separable)
Scup (with MAFMC)	VPA (ADAPT)
Sea herring	VPA (ADAPT, ICA)
Shad and river herring	Stock recruitment
Spanish mackerel	VPA (ADAPT)
Spot	None
Spotted seatrout	None
Striped bass	Spawning stock biomass, stock recruitment
Summer flounder (with MAFMC)	VPA (ADAPT)
Tautog	VPA (ADAPT, Laurec-Shepard, ESA), catch curves, tagging
Weakfish	VPA (ESA)
Winter flounder	VPA (ADAPT)

NOTE: ADAPT = Adaptive approach (age-structured); ESA = extended survivors analysis; ICA = integrated catch analysis; MAFMC = Mid-Atlantic Fishery Management Council; SAFMC = South Atlantic Fishery Management Council; VPA = virtual population analysis.

8. Lack of information for stock identification may affect appropriateness of stock assessment models applied to various species.

9. There is need for fisheries-independent data in stock assessment modeling, particularly as tuning indices for VPA analyses.

10. Relatively simple models may overlook important parameters that the model is not robust to; i.e., sensitivity to changes in catchability.

11. Little data is available on the functional relations among co-occurring exploited species and effects of fishing on shifts in ecological relations.

12. Models used to forecast how management actions will change F do not always take into account economic and social factors.

13. The quality of the input data affects the precision of the assessments.

Pacific States Marine Fisheries Commission (PSMFC)

There are fisheries on the West Coast that do not have the basic population data necessary to allow adequate assessment modeling. The thresher shark fishery is one example of a fishery that is being managed very conservatively due to the lack of funding for adequate assessments. We believe it is critical that efforts be initiated to develop the databases necessary to manage these fisheries.

Rockfish are targeted by both commercial and recreational fisheries off the West Coast. There has been concern, coast-wide, regarding the management of nearshore rockfish. These concerns have been discussed on a regional basis by the Canada-U.S. Groundfish Committee. This committee listed the following concerns.

1. The lack of biological information and abundance for many nearshore rockfish species
2. The generally poor track record of rockfish management coast-wide

TABLE H.2   Stock Assessment Methods Used by the State of Oregon to Manage Its Fisheries

Species Name	Stock Assessment Model	Comments
Pink shrimp	No formal assessment. Stock status is monitored using a retrospective, area-based index of recruitment.	Managed by season and aggregate size limit
Red sea urchin	No formal assessment. Stock status is monitored using retrospective indices of size distribution and abundance.	Managed by size limit and limited entry
Dungeness crab	None	Managed by season, sex, and size limit
Pacific herring, Yaquina Bay	No formal assessment. Spawn deposition surveys conducted to establish harvest guidelines.	Managed by harvest guideline, season, limited entry

3. The notable difficulty in managing nearshore species

4. The longevity and vulnerability to over-exploitation associated with these species even when some biological parameters are known

Rockfish are managed by the Councils and the states. Most species occur in fisheries managed by different jurisdictions making it difficult to fully assess the populations and fishery impacts.

Another issue is the lack of total catch information in many fisheries. Landing tickets give managers information on those species landed, but very limited data exist for West Coast fisheries total catch. Regulatory and economic discards are virtual unknowns. We believe an effective observer program is essential for the future health of these fish stocks and their attendant fisheries. A voluntary pilot program funded by the West Coast trawl industry is an important start to gather this data. Two states in the region provided information about the stock assessment methods they use: Oregon (Table H.2) and Alaska (Tables H.3 to H.5).

Alaska Department of Fish and Game (ADF&G)

1. Often we fail to recognize that most samples taken for stock assessment are not random samples. Statistical models behind estimates from stock assessment are built with the concept of samples being representative of populations because these samples were randomly drawn. In stock assessment, samples are almost never randomly drawn. More diagnostic testing should be built into stock assessments to determine if our non-randomly drawn samples are, or are not, representative.

2. Extrapolation of results from VPA (virtual populations analysis) to manage fisheries in the current year is a poor substitute for annual surveys and mark-recapture experiments. Since these extrapolations reflect past, not current information, results of VPA lag behind actual trends in abundance. Ancillary information from indices such as mean CPUE (catch per unit effort) to compensate adds another layer of assumptions and considerable danger.

3. Sampling (handling) fish affects their subsequent behavior in ways that could bias stock assessments based on that behavior. Again, more diagnostics are needed to determine influences of handling.

4. The lack of age structures for most shellfish species has thwarted the development of assessment models for years. For the past three years, ADF&G has focused on development of length-based methods for shellfish stock assessments under a range of data situations (e.g., trawl survey, pot survey, no survey). During the next few years ADF&G will continue to apply these methods to various stocks around the state. ADF&G also plans to

TABLE H.3   Methods Used to Assess Alaska Shellfish

Species Name/Area	Stock Assessment Model	Comments
Red king crab		
Southeast Alaska	Catch-survey analysis (CSA)	For CSA methods, see Collie and Kruse (in press)
Bristol Bay	Length-based analysis (LBA)	For LBA methods, see Zheng et al. (1995a,b)
Kodiak	CSA, trawl survey, area-swept estimator	Plans: LBA
Prince William Sound, Cook Inlet, Alaska Peninsula, Pribilof Islands, Norton Sound.	Trawl survey, area-swept estimator	Plans: CSA (historical pot surveys) and LBA (ongoing trawl surveys)
Adak	None	Plans: catch-length analysis (CLA) - see Zheng et al. (1996) for methods
Blue king crab		
St. Matthew and Pribilof Islands	Trawl survey, area-swept estimator	CSA in progress
Prince William Sound	None	CPUE index from experimental fishery
Golden king crab		
Southeast Alaska, Prince William Sound, Dutch Harbor, Adak	None	Plans: GIS analysis of fisheries with on-board observer data
Tanner crab		
Southeast Alaska	None	CPUE model used to project attainment of fixed quota
Cook Inlet, Kodiak, Alaska Peninsula, Bering Sea	Trawl survey, area-swept estimator	LBA in progress for Bering Sea, LBA planned for Kodiak
Snow crab		
Bering Sea	Trawl survey, area-swept estimator	Needs: growth and *M* estimates
Hair crab		
Bering Sea	Trawl survey, area-swept estimator	Needs: growth and *M* estimates
Dungeness crab		
Southeast Alaska, Yakutat, Kodiak, Alaska Peninsula	None	CLA possible in future
Prince William Sound, Cook Inlet	Pot survey, index of abundance	CSA possible in future
Pink shrimp		
Cook Inlet, Kodiak, Alaska Peninsula	Trawl survey, area-swept estimator	LBA in progress for Cook Inlet
Southeast Alaska	None	CLA possible in future
Weathervane scallop		
Prince William Sound, Cook Inlet	Dredge survey, area-swept estimator	Plans: develop age-structured model
Yakutat, Kodiak, Dutch Harbor, Bering Sea	None	Plans: GIS analysis of onboard fishery observer data
Sea cucumber		
Southeast Alaska	Dive survey, surplus production model	Needs: growth and recruitment estimates
Elsewhere	None	Needs: sampling, basic biological information
Sea urchin		
Southeast Alaska	Dive survey, surplus production model	Needs: growth and recruitment estimates
Elsewhere	None	Needs: sampling, basic biological information
Intertidal clams		
Cook Inlet (Kachemak Bay)	Transect surveys	Plans: develop age-structured model
Elsewhere	None	
Miscellaneous other shellfish species (crabs, shrimps, clams, etc.)		
Statewide	None	Needs: basic biological information and sampling programs

develop fishery-based assessments for fisheries with onboard observers in a GIS framework. Despite progress, stock assessments are not possible for the majority of Alaskan shellfish stocks due to a lack of basic biological information and funds for sampling programs.

5. Budgets are declining while the complexity of and conflicts with the fishery management process are increasing. This situation will inevitably lead to a reduction in the quality of stock assessment data. This problem is most acute for stocks where fisheries are developing.

Gulf States Marine Fisheries Commission (GSMFC)

The stock assessment team of the Gulf States Marine Fisheries Commission has reviewed the documents provided by your office and generally offer complete concurrence with the thirteen points raised by the Atlantic States staff. Three comments (which to me represent emphasis) forwarded by GSMFC stock assessors were:

1. Environmental Concerns. Natural climatological phenomena obviously play a role in success/failure of fishery management measures. While this is taken for granted in a functional sense, there is no practical way to account for these influences in current modeling practices. (Related or similar to ASMFC #5.)

2. Human Dimensions. Almost totally lacking in stock assessment proceedings is the reactive capability of the fishers. It is typical for fishers to act or react unpredictably to management measures. (Related or similar to ASMFC #12.)

3. Local vs. Regional Management. Spotted seatrout is an example. While regional management is desirable from an interstate transport perspective, status of local trout stocks can be drastically different within a relatively small geographic region. Regional managers may know very little about the condition of these local stocks. (Imbedded in ASMFC #s, 2, 5, 6, 8, 9, 10, and 13.)

TABLE H.4   Methods Used to Assess Pacific Salmon

Species Name/Area	Harvest Policy	Stock Assessment Methods
Chinook salmon		
Southeast Alaska	Preseason quota based on constant harvest rate applied to preseason forecasted abundance with Pacific Salmon Commission Chinook Technical model	Cohort analysis of coded wire tag recoveries from hatchery indicator stocks; wild stock escapement enumeration with weirs, mark-recapture, aerial surveys, foot surveys
Yakutat	Fixed escapement goal	Weir counts
Copper River	Fixed escapement goal	Aerial survey counts
Upper Cook Inlet	Limited directed commercial fishing, reduce bycatch, sport fisheries managed for fixed escapement goals	Sonar counts, weir counts, aerial surveys
Kodiak/Chignik	No directed commercial fishing, fixed escapement goal	Aerial surveys, weir count
Bristol Bay	Fixed escapement goal	Sonar counts, aerial surveys
Kuskokwim River	Commercial fishery quota ranges	Test fishery catches, commercial fishery CPUE
Yukon River	Commercial fishery quota ranges	Test fishery catches, commercial fishery CPUE, mark or recapture, postseason run reconstruction
Coho salmon		
Southeast Alaska	Scheduled closures of troll fishery based on CPUE of inside fisheries and early escapement; postseason evaluation of escapement relative to goals and harvest rates for indicator stocks	Fishery CPUE, indicator stock assessments (weir counts, mark-recapture of juveniles, exploitation rates CWT marking), fishwheel catches, aerial surveys
Yakutat	Fixed escapement goals	Weir counts, aerial surveys
Copper, Bering River	Fixed escapement goals	Weir counts
Upper Cook Inlet	Stocks in Northern/Central District systems caught incidental to directed sockeye fishery. Directed Westside set net fishery limited to two openings per week and reduced if CPUE low	Fishery CPUE, limited weir count
Kodiak	Fixed escapement goal	Aerial survey, weir count
Chignik	Fixed escapement goal	Aerial survey, weir count
Alaska Peninsula	Fixed escapement goal	Aerial survey, weir count
Bristol Bay	Fixed escapement goal	Sonar count, aerial survey
Kuskokwim River	Reduced fishing periods when run strength is weak	Test fishery CPUE
Yukon River	Catch incidental to fall chum fishery, stock not fully exploited	Test fishery CPUE, sonar count
Pink salmon		
Southeast Alaska	Fixed escapement goals	Aerial survey
Prince William Sound	Fixed escapement goals	Aerial survey, hatchery stock identification based on CWT
Lower Cook Inlet	Fixed escapement goals	Aerial survey
Kodiak	Fixed escapement goals	Aerial survey
Chignik	Fixed escapement goals	Aerial survey
Alaska Peninsula	Fixed escapement goals	Aerial survey
Bristol Bay, Norton Sound	Production highly variable, markets not well developed	None
Chum salmon		
Southeast Alaska	Generally catches incidental to pink salmon, except for hatchery terminal harvest	Aerial surveys, poor quality due to presence of pink salmon
Prince William Sound	Generally catches incidental to pink salmon, except for hatchery terminal harvests	Aerial surveys, poor quality due to presence of pink salmon
Kodiak	Generally catches incidental to pink salmon, except for hatchery terminal harvests	Aerial surveys, poor quality due to presence of pink salmon

TABLE H.4   (Continued)

Species Name/Area	Harvest Policy	Stock Assessment Methods
Alaska Peninsula	Generally catches incidental to pink salmon	Aerial surveys, poor quality due to presence of pink salmon
Bristol Bay	Harvests incidental to directed sockeye fishery	Nushagak sonar, aerial survey
Kuskokwim River	Fixed escapement goal	Sonar count, test fishery catches, weir counts, aerial survey
Yukon River	Fixed escapement goals, in river allocation plan, subsistence priority summer run fishery is market limited	Sonar count, test fishery catches
Norton Sound	Fixed escapement goal, large-run fisheries are market limited	Lower counts, aerial surveys, weir counts
Kotzebue Sound	Fixed escapement goal	Sonar count, aerial survey, test fishery catches
Sockeye salmon		
Southeast Alaska	Fixed escapement goals, interception fishery limited by quota	Weir count, mark-recapture
Yakutat	Fixed escapement goal	Weir count, aerial survey
Copper/Bering River	Fixed escapement goal	Sonar count, aerial survey
Upper Cook Inlet	Fixed escapement goal	Sonar count, test fishery catches
Kodiak	Fixed escapement goal	Weir count, test fishery catches
Chignik	Fixed escapement goal	Weir count, test fishery catches
Alaska Peninsula	Fixed escapement goal, June interception quota set on Bristol Bay sockeye preseason forecast and chum cap	Weir count, aerial survey, test fishery catches
Bristol Bay	Fixed escapement goal	Lower count, sonar count, aerial survey, test fishery catches

TABLE H.5 Methods Used to Assess Pacific Herring

Area	Harvest Policy	Stock Assessment Methods
Southeast Alaska	Fixed harvest rate with threshold	Age-structured analysis tuned to biomass estimated with spawn deposition surveys
Prince William Sound	Fixed harvest rate with threshold	Age-structured analysis tuned to biomass estimated with spawn deposition surveys, aerial surveys
Lower Cook Inlet	Quota based on historical catches that maintained stable abundances; quota increments or decrements based on in-season abundance indices	Aerial surveys
Kodiak	Quota based on historical catches that maintained stable abundances; quota increments or decrements based on in-season abundance indices	Aerial surveys
Bristol Bay	Fixed harvest rate with threshold	Aerial surveys
Kuskokwim Bay	Quota based on historical catches that maintained stable abundances; quota increments or decrements based on in-season abundance indices	Aerial surveys
Norton Sound	Quota based on historical catches that maintained stable abundances; quota increments or decrements based on in-season abundance indices	Aerial surveys

TABLE H.6   Stock Assessment Methods Used by the Gulf States Marine Fisheries Commission

Species Name	Stock Assessment Methods	Comments
Gulf sturgeon	None—coast-wide annual population estimate in some Florida rivers	Listed as threatened; habitat and population status largely unknown
Striped mullet	VPA (GXPOPS, Florida) SSBR, SPR coast-wide	
Gulf menhaden	VPA (basic and separable)	
Black drum	VPA Length cohort analysis	
Striped bass	Limited SSBR Tagging studies	
Oyster	None	Local annual predictive models
Blue crab	None	Stock assessment to be attempted for 1997 FMP revision
Spanish mackerel	Surplus production yield per recruit VPA—Florida	
Gulf shrimp	Indices of abundance	
Spotted seatrout	VPA SPR	FMP to be published in 1997
Flounder	In progress (VPA, SPR)	FMP to be published in 1997

NOTE: FMP = fishery management plan; SPR = spawning biomass per recruit; SSBR = spawning stock biomass per recruit; VPA = virtual population analysis.

I

Model[*] Results in Terms of Exploitable and Total Biomass

---

*The following abbreviations are used for models: ADAPT = Adaptive approach (age-structured); ADMB = Autodifferentiation Model Builder; ASA = Age-structured assessment models; SS = Stock Synthesis.

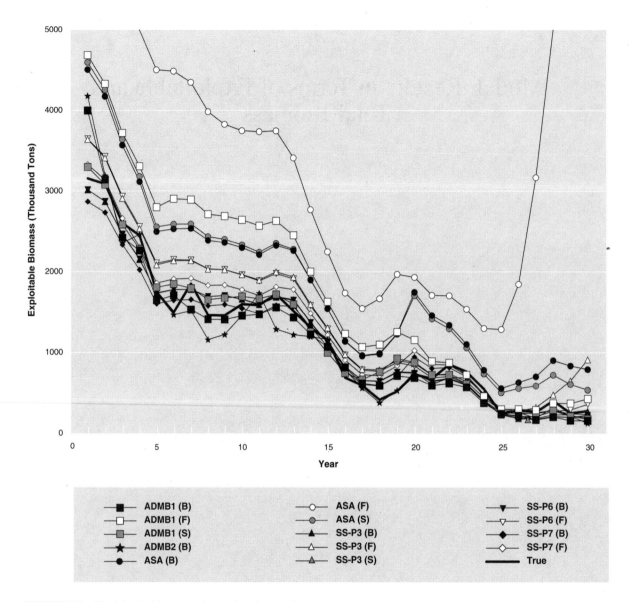

FIGURE I.1   Exploitable biomass values using data set 1.

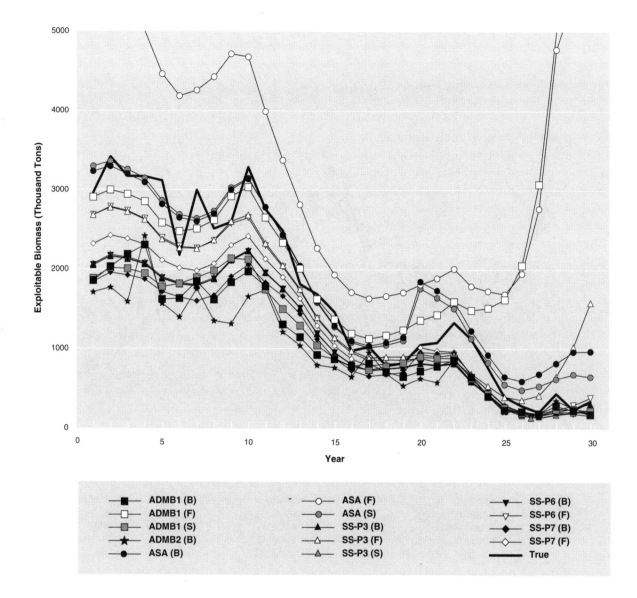

FIGURE I.2   Exploitable biomass values using data set 2.

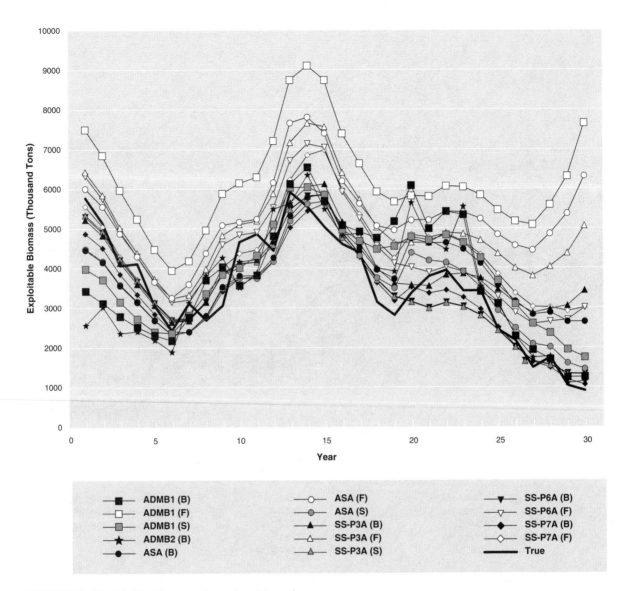

FIGURE I.3 Exploitable biomass values using data set 3.

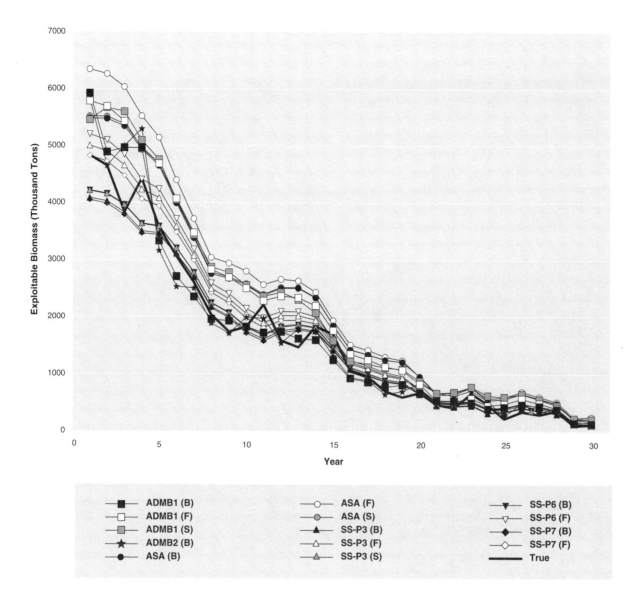

FIGURE I.4 Exploitable biomass values using data set 4.

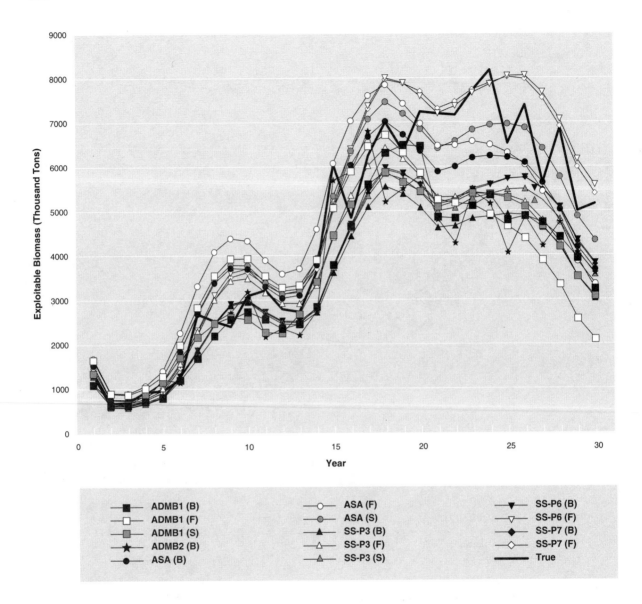

FIGURE I.5   Exploitable biomass values using data set 5.

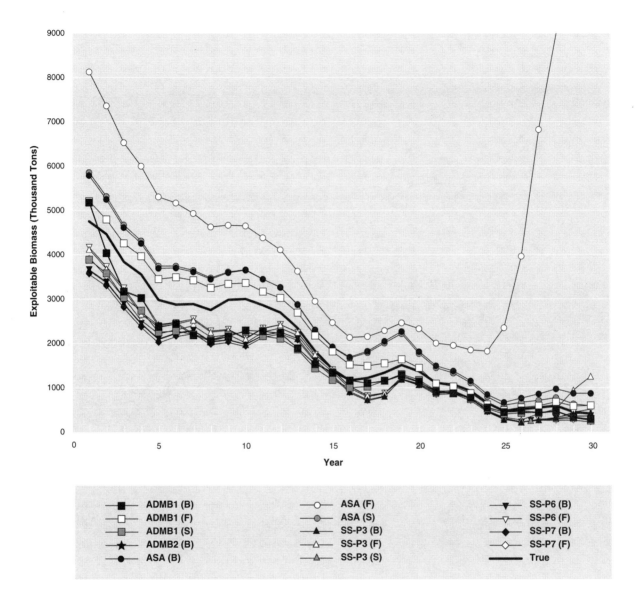

FIGURE I.6   Total biomass values using data set 1.

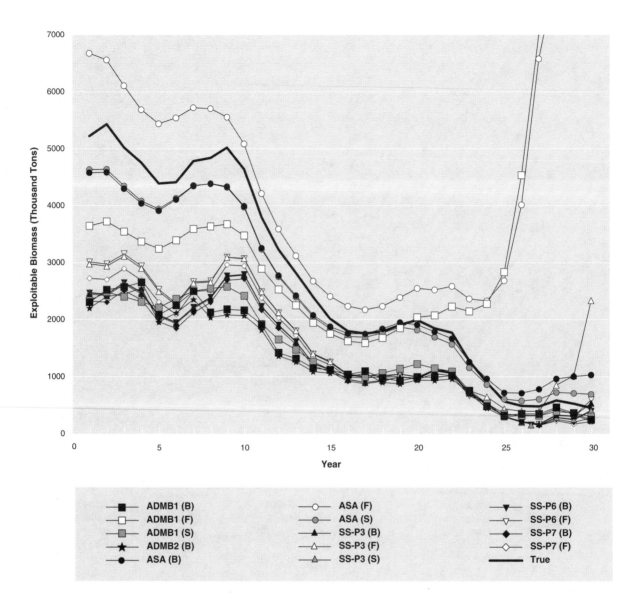

FIGURE I.7   Total biomass values using data set 2.

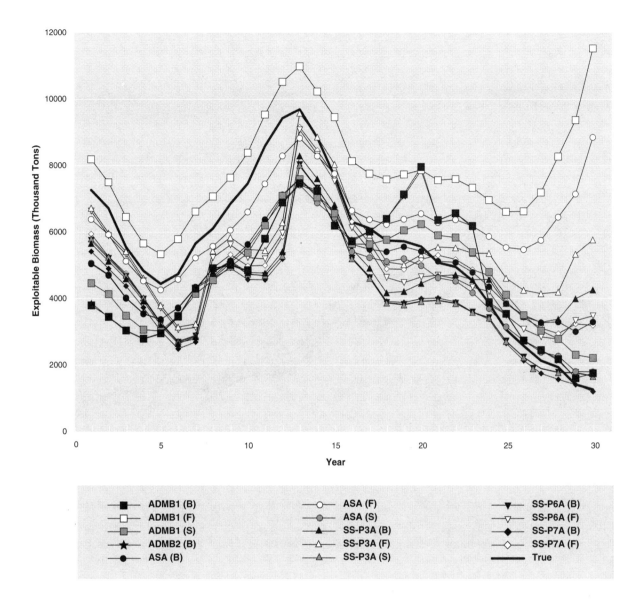

FIGURE I.8   Total biomass values using data set 3.

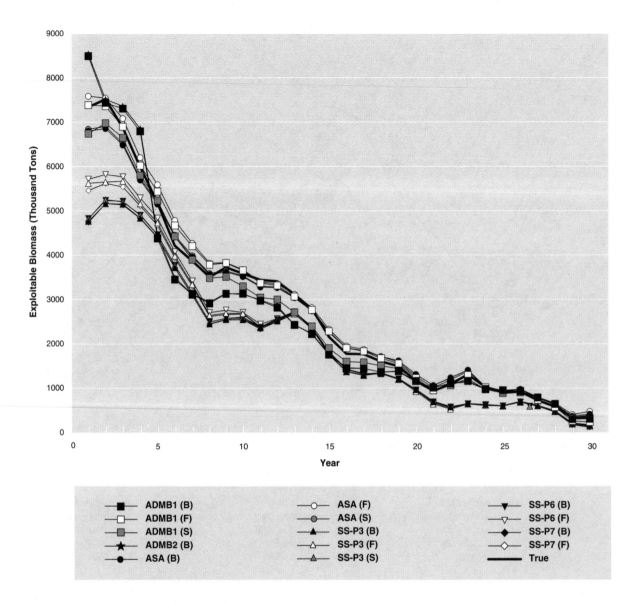

FIGURE I.9   Total biomass values using data set 4.

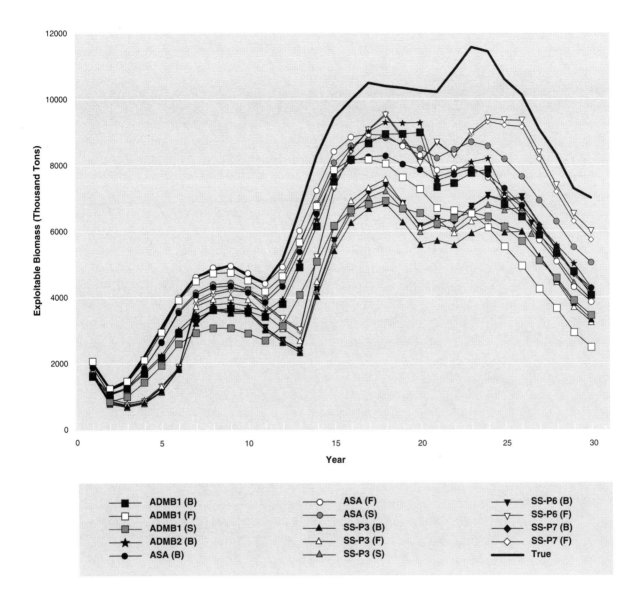

FIGURE I.10   Total biomass values using data set 5.

J

Excerpts from the Food and Agriculture Organization Report *Precautionary Approach to Fisheries**

4.3 ASSESSMENT METHODS AND ANALYSIS

69. Biological reference points for overfishing should be included as part of a precautionary approach.

70. A precautionary approach specifically requires a more comprehensive treatment of uncertainty than is the current norm in fishery assessment. This requires recognition of gaps in knowledge, and the explicit identification of the range of interpretations that is reasonable given the present information.

71. The use of complementary sources of fishery information should be facilitated by active compilation and scientific analysis of the relevant traditional knowledge. This should be accompanied by the development of methods by which this information can be used to develop management advice.

72. Specifically the assessment process should include:

a. scientific standards of evidence (objective, verifiable and potentially replicable) should be applied in the evaluation of information used in analysis;
b. a process for assessment and analysis that is transparent; and
c. periodic, independent, objective and in-depth peer review as a quality assurance.

73. A precautionary approach to assessment and analysis requires a realistic appraisal of the range of outcomes possible under fishing and the chances of these outcomes under different management actions. The precautionary approach to assessment would follow a process of identifying alternative possible hypotheses or states of nature, based on the information available, and examining the consequences of proposed management actions under each of these alternative hypotheses. This process would be the same in data rich and data poor analyses. A precautionary assessment would, at least, aim to consider: (a) uncertainties in data; (b) specific alternative hypotheses about underlying biological, economic and social processes, and (c) calculation of the response of the system to a range of alternative management actions. A checklist of issues for consideration under these headings is found in the following paragraphs.

---

*FAO, 1995a.

74. Sources of uncertainty in data include: (a) estimates of abundance; (b) model structure; (c) parameter values used in models; (d) future environmental conditions; (e) effectiveness of implementation of management measures; (f) future economic and social conditions; (g) future management objectives, and (h) fleet capacity and behavior.

75. Specific alternative hypotheses about underlying biological, economic and social processes to be considered include (a) depensatory recruitment or other dynamics giving rapid collapse; (b) changes in behavior of the fishing industry under regulation including changes in coastal community structure; (c) medium term changes in environmental conditions; (d) systematic under-reporting of catch data; (e) fishery dependent estimates of abundance not being proportional to abundance; (f) changes in price or cost to the fishing industry; and (g) changes in ecosystems caused by fishing.

76. In calculating (simulating) the response of the system to a range of alternative management actions, the following should be taken into account:

a. short term (1-2y) projections alone are not sufficient for precautionary assessment; time frames and discount rates appropriate to inter-generational issues should be used, and

b. scientific evaluation of management options requires specification of operational targets, constraints and decision rules. If these are not adequately specified by managers then precautionary analysis requires that assumptions be made about these specifications, and the additional uncertainty in the consequences resulting from these assumptions should be calculated. Managers should be advised that additional specification of targets, constraints and decision rules are needed to reduce this uncertainty.

77. Methods of analysis and presentation will differ with circumstances, but effective treatment of uncertainty and communication of the results is necessary in a precautionary assessment. Some approaches (see the Appendix to this section) that could prove useful are:

a. where there are no sufficient observations to assign probabilities to different states of nature that have occurred, decision tables could be used to represent different degrees of management caution through the Maximin and Minimax criteria;

b. where the number of different states of nature and the number of potential management actions considered is small, but probabilities can be assigned, then decision tables can be used to show the consequences and probabilities of all combinations of these, and

c. where the range of states of nature is large the evaluation of management procedures is more complex, requiring integration across the various sources of uncertainty.

78. A precautionary approach to analysis would examine the ability of the data collection system to detect undesirable trends. When the ability to detect trends is low, management should be cautious.

79. Since concerns regarding the reversibility of the adverse impacts of fishing are a major stimulus for a precautionary approach, research on reversibility in ecosystems is an important part of developing precautionary approaches.

K

Uncertainty in Stock Assessment Methods and Models

INTERVAL ANALYSIS

In much of fisheries science, the assignment of quantitative values to parameters is inexact. Consequently, a measurement is never known certainly, and scientists tend to report measured values as the best estimate and the possible error around the estimate. This is sometimes expressed with the "plus or minus" convention or as an interval that is believed to contain the actual value. Interval analysis, described in detail by Moore (1966, 1978) and by Alefield and Herzberger (1983), is considered useful for this type of uncertainty projection.

Briefly, interval analysis permits one to circumscribe estimates with bounds, such as $a = [a_1, a_2]$ where $a_1 < a_2$, and to use range arithmetic to propagate uncertainty. A simple example is provided by illustrating the addition of intervals. Suppose we know that M, the instantaneous natural mortality rate, is between 0.2 and 0.4 and F, the instantaneous fishing mortality rate, is between 0.5 and 0.8, and we want an estimate of Z, the total instantaneous mortality rate, as the sum of these two intervals. Interval addition proceeds as follows:

$$[M_1 M_2] + [F_1 F_2] = [M_1 + F_1], [M_2 + F_2]. \tag{K.1}$$

By substitution, we obtain

$$[0.2, 0.4] + [0.5, 0.8] = [0.7, 1.2].$$

From the above, we know that the value for Z must be in the interval [0.7, 1.2]. Rules for various other operators, such as subtraction, multiplication, and division, are relatively simple. However, interval analysis is not easy to use in complex calculations. Fortunately, software capable of handling all computational details is available (Ferson and Kuhn, 1994).

Although interval arithmetic is relatively easy to explain and to apply and seems to work regardless of the source of uncertainty, there are some shortcomings. For example, ranges can grow rapidly in some applications, so that the results may lead to overly conservative fishery management advice. In addition, there is an apparent paradox because no exact values are specified, but the end points (bounds) are presumed to be exact. In summary, interval analysis may be appropriate in situations where no repetitive sampling has occurred and no probability is defined.

FUZZY ARITHMETIC

Fuzzy numbers have been described by Ferson and Kuhn (1994) as a generalization of intervals that can serve as representations of values whose magnitudes are known with uncertainty. Fuzzy numbers can be thought of as a stack of intervals, each at a different level of presumption (alpha), which ranges from 0 to 1. The range of values is narrowest at alpha level 1, which corresponds to the optimistic assumption about the breadth of uncertainty. More formally, a fuzzy number is defined as a range of values along with a fuzzy set function in [0, 1], which is convex and reaches a maximal value of 1. The function indicates the degree to which a particular value belongs to the set.

This brief discussion is limited to fuzzy arithmetic, which is a derivation of the considerably broader field of fuzzy set theory and fuzzy logic. Fuzzy arithmetic is considered to be a part of possibility theory introduced by Zadeh (1978) and further developed by Kaufmann and Gupta (1985) and Dubois and Prade (1988). Possibility theory is somewhat analogous to probability theory, but it can be carried out under weaker assumptions and thus used when limited experimental or observational information is available. When a complete probabilistic analysis is not desired because of a lack of information about the distributions of parameters, fuzzy arithmetic may still be used to obtain reasonable, but somewhat crude, estimates of the uncertainty being propagated.

An extremely simple example of fuzzy addition follows (Figure K.1). This example adds the fuzzy numbers [B: 2000, 4000, 7000] and [A: 1000, 2000, 5000]. The actual arithmetic operations are performed only on the apex values of these triangular fuzzy numbers; thus, 4000 + 2000 = 6000. The low ranges of the two fuzzy numbers are then added together forming the base of the arithmetic result. In this case, the base of fuzzy 4000 ranges from 2000 to 7000 or 5000. The base of fuzzy 2000 ranges from 1000 to 5000, or 4000. The sum of 5000 and 4000 is 9000. The sum is divided by 2 to yield a quotient of 4500. This value is subtracted from the result of the arithmetic operation on the base number: 6000 minus 4500 in this example. Thus, 1500 becomes the left hand limit of the base. The quotient is then added to the result of the arithmetic operation, namely, 6000 + 4500 = 10500. This number is the right-hand limit of the base. The result of the addition is the fuzzy number [1500, 6000, 10500].

Limited fuzzy logic applications have already been developed in fisheries by Saila (1992), Ferson (1994), and Sakuramoto (1995). However, specific application of fuzzy arithmetic to existing fishery models has not received substantial consideration. As in the case of interval analysis, software for executing fuzzy arithmetic operations is available.[*] Using fuzzy arithmetic software minimizes the possibility of computational mistakes in complex calculations. In summary, fuzzy arithmetic is possibly an even more effective method than interval analysis for accommodating subjective uncertainty in fishery assessments; its utility should be examined more carefully for future stock assessment applications.

Because of the increasing use of Monte Carlo methods in stock assessment activities, considerably more attention should be given to determining the reliability of Monte Carlo results as a function of uncertainty in the input parameter distributions and in model assumptions. The distinction between objective and subjective uncertainties[†] should be recognized explicitly. Probability theory could be used to propagate objective uncertainty, and interval analysis or fuzzy arithmetic should be explored as an alternative to Bayesian priors for the propagation of subjective uncertainty in future stock assessment work.

---

[*]Examples include RiskCalc developed by Ferson and Kuhn (1994) and FuziCalc[TM] available from FuziWare Inc., Knoxville, Tennessee. The committee does not necessarily endorse the use of this software.

[†]According to Ferson and Ginzburg (in press), there are two basic kinds of uncertainty. The first kind, termed objective uncertainty, arises from variability in the underlying stochastic system. The second kind is called subjective (epistemic) uncertainty, which results from incomplete knowledge about a system. Probability theory seems to provide methods appropriate for projecting random variability through calculations. However, if subjective uncertainty is treated as if it were random, as in a Bayesian prior, it can result in narrower bounds on the resulting estimates than might be suitable, given that the prior itself inputs some knowledge about the parameter into the system.

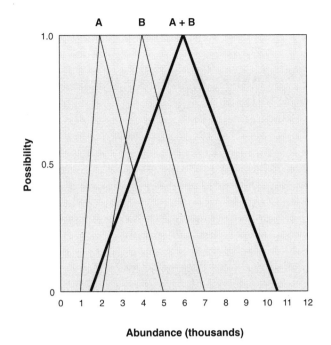

FIGURE K.1   An example of the use of fuzzy arithmetic.   The ordinate measures possibility ranging from 0 to 1, and the abscissa shows abundance (in thousands) for the example illustrated in the text.   The addition of fuzzy numbers corresponds to the addition of intervals at each possibility level from 0 to 1.